Swarm Minds

How Simple Creatures Solve Complex Problems Together

Team Trex

Published by NobleTrex Press

For permissions and other inquiries, write to:
P.O. Box 3132, Framingham, MA 01701, USA

Contents

Introduction

Imagine, for a moment, that you are blind. You are also deaf, and your sense of smell is limited to the patch of ground immediately beneath your feet. You possess no memory of the past few minutes, and you certainly have no concept of the future. You are hungry, vulnerable, and alone in a landscape that is vast and terrifyingly complex.

Now, imagine I ask you to build a cathedral.

Not just a pile of mud, but a climate-controlled architectural marvel with ventilation shafts, brooding chambers, and defensive fortifications. And I want you to do it without a blueprint, without an architect, and without speaking a word to the millions of other blind, deaf builders working alongside you.

By all logical standards, this should be impossible. It sounds like the setup for a cruel philosophical joke. Yet, in the red soil of the African savanna, millions of termites-creatures with brains the size of a grain of sand-accomplish this engineering feat every day. They build mounds that, relative to their body size, rival our skyscrapers. They maintain internal temperatures within a fraction of a degree while the outside air swings from baking heat to chilling cold. They do this not because they are smart. In fact, if you place a single termite on a table, it is a helpless, wandering speck that will quickly die.

They do it because they are part of a swarm.

This book is an exploration of a phenomenon that feels dangerously like magic until you look under the hood. It is the story of how stupidity, when multiplied and networked, becomes brilliance. It is the study of *collective intelligence*, a field that overturns our most cherished assumptions about what it means to be smart, what it means to lead, and what it means to be an individual.

For centuries, human observers looked at the synchronized aerial ballet of a starling flock or the disciplined march of army ants and assumed there must be a conductor. We projected our own hierarchical societies onto nature. We assumed the "Queen" bee was issuing orders, or that the lead bird in a V-formation was the wisest navigator. We were wrong.

As we have peeled back the layers of biological complexity, we have discovered something far more profound. There is no conductor. There is no general. There is only the hum of a million tiny interactions, a silent language of bumps, chemical trails, and flashes of light that allows order to erupt spontaneously from chaos. This is the science of the "superorganism," where the group possesses cognitive abilities that transcend the members that compose it.

Why does this matter to us? We, the big-brained primates who invented jazz, geometry, and the internet?

It matters because we are realizing that the "swarm" is not just a curiosity of the natural world; it is the fundamental architecture of complexity itself. Your own brain is, in essence, a swarm. A single neuron is as blind and limited as that termite. It knows nothing of Shakespeare or quantum mechanics; it only knows whether to fire an electrical impulse or stay silent. Yet, when eighty-six billion of them network together, the result is *you*.

Furthermore, we are living in an era where our own systems are becoming too complex for top-down control. Our power grids, our financial markets, and our traffic systems behave more like living flocks than rigid machines. To manage the future, we are turning to the algorithms of the wild. Computer scientists are studying ant colonies to route internet traffic. Roboticists are looking to fish schools to design fleets of underwater drones. We are standing on the precipice of creating our own silicon swarms, and to do so safely, we must first understand the biological software that has been refined over millions of years of evolution.

Our exploration begins with the shattering of the "Great Man" theory of nature. We often find it unsettling to accept that complex patterns do not require a master planner. It is intuitive to think that a beautiful design requires a designer. But as we will see, nature operates on a different principle: *emergence*. This is the mathematical alchemy that allows simple rules-like "stay close to your neighbor but not too close"-to generate the swirling, liquid artistry of a school of fish. We will look at how order arises from noise, and how a system with no head can survive the loss of any of its parts.

But how do these parts talk to one another? If there is no leader shouting commands, how does the group know what to do? The answer lies in a strange and silent vocabulary. For many of the creatures we will meet, the world itself is a notepad. We will delve into the concept of *stigmergy*, a term coined by French biologist Pierre-Paul Grassé in the 1950s. He observed that termites were stimulated to build by the work they had already done. The construction itself was the instruction. We will trace the invisible chemical highways laid down by ants-external memories written in pheromones that allow a colony to solve the Traveling Salesperson Problem more efficiently than some supercomputers.

This reliance on external cues leads to one of the most fascinating aspects of swarm life: the necessity of forgetting. In a world of constant change, old information is dangerous. We will see why the evaporation of a scent trail is just as important as its creation, teaching us that in dynamic systems, memory must be balanced with a healthy dose of amnesia to prevent the collective mind from getting stuck in the past.

Perhaps the most startling revelation of swarm science is that these "mindless" groups are often better at making decisions than individuals. We tend to view democracy as a messy human invention, but honeybees perfected it long before we climbed down from the trees. When a hive outgrows its home, it must find a new one. This is a life-or-death decision. If they choose a cavity that is too small, they will freeze; too large, and they cannot defend it.

We will witness the "dance floor debate," where scout bees return to the hive and perform a waggle dance to advertise potential real estate. It is a competition of ideas, purely meritocratic, where the enthusiasm of the dancer convinces others to check out the site. Through a mechanism known as *quorum sensing*, the hive shifts from debate to consensus, making a unified choice that is statistically superior to the judgment of any solitary expert. It is a masterclass in how to aggregate information without succumbing to the madness of crowds.

Of course, not all swarms are bound to the earth. We will take to the skies and the seas to understand the physics of the flock. For a long time, the instantaneous turn of a bird flock-where thousands of individuals change direction in a fraction of a second-was thought to involve telepathy. The reality is simpler and more elegant. We will explore the work of computer graphics pioneers like Craig Reynolds, who showed that three simple rules-separation, alignment, and cohesion-could simulate the breathtaking complexity of flocking on a computer screen. We will see how information travels through a school of fish like a shockwave, a mathematical phase transition that allows the group to evade predators with a reaction time faster than any single fish could muster.

But the swarm is not infallible. There is a dark side to collective feedback. The same mechanisms that allow ants to find the shortest path to food can also lock them into a loop of death. We will encounter the "ant mill," a horrific phenomenon where army ants, having lost the pheromone trail of the main group, latch onto the scent of the ant in front of them. They begin to march in a circle, a swirling vortex of blind obedience that continues until the ants die of exhaustion. It is a chilling reminder of the fragility of systems that rely on positive feedback without checks and balances-a lesson that applies uncomfortably well to financial bubbles and viral misinformation in human social networks.

As we move deeper into the mechanics of these biological machines, we will look at the cartographers of scent and the builders without blueprints. We will see army ants that do not just walk on the ground but *become* the ground, linking their bodies together to form bridges and rafts, dissolving and reforming their infrastructure based on the traffic flow. We will analyze the algorithms of mud that allow termites to build their breathing towers, structures that are essentially external lungs for the colony.

Finally, we will turn our gaze to the future. The principles of the swarm are moving from the petri dish to the microchip. Engineers are currently designing swarms of nanobots intended to flow through the human bloodstream, hunting tumors like a pack of wolves. Agricultural researchers are developing fleets of micro-drones to pollinate crops in a world where biological bees are dying out. We are building silicon mimics that operate on the laws of the jungle.

However, as we attempt to harness this power, we must ask ourselves: are we ready to share the world with alien intelligences of our own making? A swarm does not think like a human. It does not have a center. It does not have a "self" in the way we understand it. It is resilient, relentless, and often totally opaque to outside observers.

Understanding the logic of the swarm is no longer just a biological curiosity; it is a necessity for the technological age.

This book is not just a catalog of animal behaviors. It is a lens through which to view the world. Once you understand the rules of the swarm, you will never look at a crowded sidewalk, a flashing school of neon tetras, or even the firing of neurons in your own mind the same way again. You will stop looking for the leader, and start seeing the web.

So, let us step away from our human obsession with hierarchy and control. Let us descend into the undergrowth and dive into the deep ocean. Let us learn how the many become the one. Welcome to the mind of the swarm.

Chapter 1

The Invisible Conductor

When we observe the synchronized banking of a starling flock or the precise architecture of a termite mound, instinct compels us to look for a commander issuing orders. This chapter peels back the illusion of central control to reveal how order arises spontaneously from the bottom up. We will discover how immediate neighbors and simple constraints, not grand designs, orchestrate nature's most complex performances.

1.1 The Tyranny of the Immediate

Imagine standing on a hill above a murmuration of starlings at dusk. From that height the flock looks like a single, purposeful creature: it stretches, folds, and pours across the sky with the smoothness of ink in water. Patterns seem to pass through it—dark waves, sudden ripples, synchronized turns—as if some invisible conductor were cueing each movement.

Now shrink yourself into one bird.

Your world is no longer the whole flock. It is a moving bubble of air, a few beating wings close by, a sliver of horizon, a background of shifting silhouettes. You have no balcony seat. You do not know what shape the flock makes. You do not know whether the group is "escaping" or "searching" or "showing off". You know only what your senses can reach before the scene changes again.

This mismatch between what the group *looks like* from the outside and what any individual can *know* from the inside is the first hard constraint of swarm minds. It can feel almost insulting to human intuition. When people coordinate, they often do it by sharing a plan: a meeting agenda, a route, a leader's instructions, a map. We are used to explaining organization by appealing to something like overview and intention. In swarms, overview is exactly what no one has.

Biologists have a name for an animal's subjective world: *Umwelt*, a term associated with the early twentieth-century biologist Jakob von Uexküll. The idea is simple but easy to forget. Each creature lives in a sensor-built reality, stitched together from what it can detect and what it can ignore. A tick's world is famously sparse: a whiff of butyric acid, a change in warmth, the feel of hair. A fish's world is crowded with pressure changes and vibrations. A starling's world is made of contrast, motion, and distance measured in wingbeats. The flock we admire from the hillside is not the flock any bird can perceive.

The tyranny is not just that information is local. It is that information is *immediate*. In a crowd, the past dissolves quickly: the neighbor you adjusted to a moment ago is no longer in the same place; the gap you avoided has been replaced by another. There is rarely time to store a rich internal model, even if an animal had the memory or attention to do it. The swarm runs on what is available now.

A useful way to picture this is to separate two viewpoints. One is the "God's-eye" view: the camera above the flock, the data set reconstructed by a scientist, the image in a textbook. The other is the "agent" view: what one bird, one fish, one ant can sense at the scale of its own body and reaction time. The distance between these viewpoints is where most myths about swarm intelligence are born.

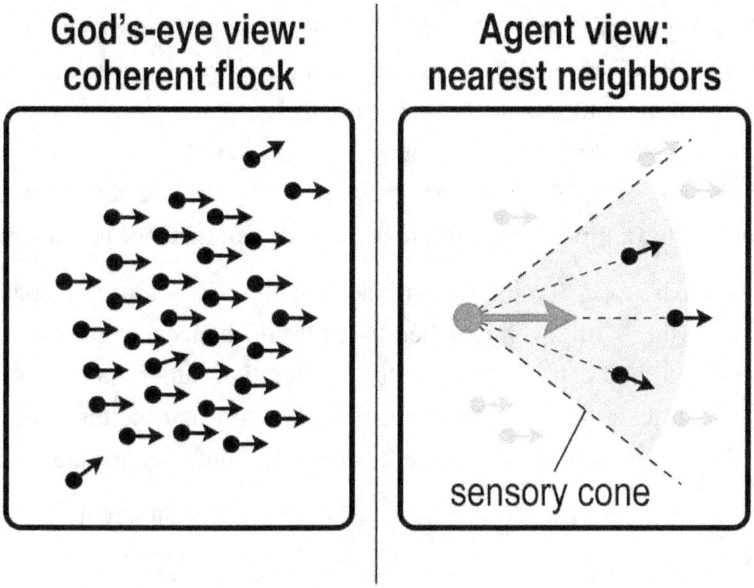

The picture is stylized, but the point is stubbornly real. Even in an animal as visually capable as a bird, the effective neighborhood is limited. You might assume a starling pays attention to everyone within a certain distance—a fixed radius like a bubble around the body. Yet field work on starling flocks, including reconstruction of three-dimensional positions by Ballerini and colleagues (2008), suggested something both stranger and more plausible: each bird seems to coordinate with a *fixed number* of neighbors (around six or seven), not all birds within a fixed metric range. This is called a *topological* interaction rule: your neighborhood is defined by who is nearest, not how many meters away they are.

Why would that matter? Picture a flock that suddenly compresses because a hawk appears. If each bird obeyed a fixed-distance rule, the number of neighbors inside that radius would spike. The bird's "social load" would change abruptly right when it can least afford distraction. A topological rule is calmer. In a dense patch or a sparse patch, the bird still tracks roughly the same number of companions. Cohesion survives big swings in density because the rule does not require the animal to recalibrate the meaning of distance under stress.

This does not make the starling clever in a human sense. It makes the starling's rule well matched to its circumstances. A topological neighborhood can be implemented by attention rather than measurement: track a handful of near bodies, the ones that matter most for collision and alignment, and let the rest blur into background.

The same constraint appears across very different creatures. A schooling fish does not consult the entire school; it responds to a few nearby fish plus the flow patterns the school itself creates. An ant on a trail does not know the trail's destination; it detects pheromone on the ground and the movement of other ants in front of it. A honeybee in a swarm does not carry the swarm's map; it feels contacts, odors, vibrations.

Local information is not merely limited; it is *noisy*. A gust tilts your neighbor. Sun glare erases contrast. A wingbeat causes a momentary wobble. If you are an individual animal, your measurements are shot through with error. And yet a flock can turn with startling crispness. How can anything coherent be built on such shaky readings?

Part of the answer is that swarms outsource stability to geometry. If you average over a handful of neighbors, noise cancels. If you keep a preferred distance, collisions become rare. If you align roughly, the group's motion becomes predictable enough that small errors do not explode. Swarms do not need perfect information; they need information that is good enough when combined with constraints that dampen mistakes. This is where *negative feedback* quietly does its work: a deviation creates a correction that opposes the deviation. Drift too close and you steer away; fall behind and the flow carries you forward; turn too sharply and you lose alignment cues, nudging you back.

The result can look, to an outside observer, like foresight. But the foresight is in the physics and the rule, not in a mental model of the whole.

A revealing historical moment came not from biology but from computer graphics. In 1987, Craig Reynolds introduced "Boids," a simulation meant to create lifelike flocking on screen. The ingredients were disarmingly plain: avoid crowding, steer toward average heading, steer toward average position. No bird knew the shape of the flock. No bird was assigned a leader role. Yet on a computer monitor, coherent flocking emerged. It was an animator's tool, but it functioned as a thought experiment: if something as visually persuasive as flocking can arise from local rules, then the appearance of choreography cannot be taken as evidence of a choreographer.

What makes this more than a cute demo is that similar minimal models in physics and mathematics had already been showing the same lesson. Tamás Vicsek and András Czirók (1995) studied a stripped-down system of self-driven particles: each particle adjusts its direction toward the average direction of its local neighbors, with some randomness added. When the noise is low enough and the density high enough, the system abruptly shifts from disordered milling to collective motion: a phase transition in the language of statistical physics. Again, no particle has global information. The "decision" to move together is not made anywhere; it is a property of the interacting whole.

The temptation, at this point, is to replace the idea of a leader with an even more mystical idea: the flock as a super-organism that somehow "knows." But the tyranny of the immediate still applies. The group's responsiveness must be built from the propagation of small, local influences: one bird adjusts to a neighbor, whose neighbor adjusts to another, and so on. That sounds slow, like a rumor spreading in a crowd. Yet real flocks can react fast enough to avoid predators. The trick is that influence can spread without any individual transmitting a message in the ordinary sense.

Empirical work by Andrea Cavagna and collaborators (2009) on starling flocks measured something subtle: correlations in the *fluctuations* of birds' velocities. Even if all birds are generally moving in the same direction, each bird's speed and direction wiggle slightly around the average. Cavagna's group found that these wiggles can be correlated over distances that scale with the flock's size—a kind of scale-free correlation. Translated into plain language, a perturbation can percolate through the group without a broadcast signal, because the system is poised so that local adjustments are strongly linked.

This begins to explain how a swarm can feel "wired together" while each animal remains myopic. The wiring is not made of messages; it is made of coupled reactions. The same is true in human crowds. In a packed subway car, you do not need to see the person at the far door to be affected by them. A shove propagates through bodies like a wave through springs. No one transmits a plan. The coupling itself carries influence.

Still, the romance of the flock can mislead us into assuming local information is always visual and always symmetric. Many animals live in sensory worlds where information arrives in lopsided ways. A fish's lateral line senses pressure differences strongly to the side. An ant's chemical senses are often best on the ground where the antennae can tap and sweep. Even in birds, vision has blind spots, and attention has limits. An animal does not respond to everything it can detect; it responds to what it can detect *and* prioritize in time.

This is why "sensory radius" is an idea that should be handled with care. It is not a neat circle drawn around an agent. It is a shifting, anisotropic, attention-dependent region shaped by anatomy, environment, and urgency. Under calm conditions, a bird may monitor more neighbors; under attack, it may simplify. Under low light, it may rely more on motion cues than on detailed position. The neighborhood is a moving compromise between what is available and what is affordable.

There is also a darker implication. If local information is all you have, you are vulnerable to local deception. Humans know this from everyday life: a rumor that looks convincing in your immediate social circle can be false at the scale of a city. Swarms face analogous problems. An ant's pheromone trail can be hijacked by parasites or disrupted by rain. A fish can be misled by a sudden flash or turbulence. A bird can be pulled off course by a few neighbors panicking early. The swarm's intelligence is powerful, but it is not omniscient. It is bounded by the quality of local cues and by the possibility that short-range rules can amplify the wrong thing.

A concrete example comes from the world of ants, where the apparent wisdom of the colony is often nothing more than an amplified accident. In classic trail formation, a few ants leave the nest and wander. If one finds food and returns, it deposits pheromone along its path. That chemical trace slightly increases the chance that another ant will follow the same route. If that second ant also finds food and returns, it reinforces the same trace. This is *positive feedback*: the strong get stronger. Soon a bustling trail forms that looks like deliberate planning, even though it began with a stumble in the right direction.

From the ant's point of view, the world is brutally immediate. "Strong smell here; keep going." "Weak smell; turn." "Bump into another ant; adjust." The ant does not need to know that a trail is "optimal" or that the food source will run out. Indeed, this is why ant trails can become stubborn. If the environment changes, the colony may continue to exploit a now-inferior path because the local signal remains strong. Local rules can lock a group into a local maximum, in the language of optimization: good enough nearby, wrong globally.

Bird flocks have their own version of this trap. A sudden turn by a few individuals can cascade, not because the turn is correct, but because the coupling is strong and the cost of hesitation is high. Under predation, the safest bet may be to turn with your neighbors regardless of whether you personally saw the hawk. In that sense, a flock's spectacular unity is not always an achievement of knowledge; it can be an achievement of *trust* under uncertainty, engineered by evolution because distrust is too slow.

The tyranny of the immediate also reshapes what "communication" means. Human communication is often symbolic: words that refer to things not present, plans for later, counterfactuals. Swarm communication is typically indexical and embodied: a chemical concentration, a vibration, a visual alignment cue. It refers to the here-and-now state of the local world. This is why swarms can be fast. They do not translate experience into abstract sentences. They push information directly into action.

There is a humility in this perspective. The organized pattern we admire from above is not a plan executed; it is a constraint satisfied. Each agent follows rules that are feasible with its sensors and reflexes. Each agent is trapped, in the best possible way, inside the present moment. Out of that trap comes a kind of collective grace: not because anyone escapes local ignorance, but because local ignorance is the only raw material the system ever had.

1.2 Alchemy of the Crowd

A single ant can be an underwhelming creature. It walks, stops, turns, taps the ground with its antennae, and walks again. If it meets another ant it may pause for a fraction of a second, as if exchanging a brief handshake. No one watching at ground level would call it strategic. And yet, scaled up to thousands of ants, the colony can look like an engineer: it lays highways, allocates labor, regulates food intake, and responds to disaster with a speed that makes human institutions seem sluggish.

The uneasy feeling comes from the mismatch between cause and effect. How can such plain ingredients yield something that looks, from the outside, like planning?

The word most often used for this transformation is *emergence*. It is sometimes spoken with a mystical tone, but it need not be mystical at all. Emergence simply means that when many parts follow rules of interaction, the group can display regularities that are not properties of any single part. A water molecule is not wet; a crowd can be a stampede even though no person intends it. The important move is not to treat the group pattern as a separate ghostly agent, but to treat it as a predictable outcome of amplification, constraint, and feedback.

Feedback is the quiet machinery behind much of swarm order. *Positive feedback* makes a tendency grow: the more a path is used, the more attractive it becomes; the more an idea is repeated, the more likely someone else repeats it. *Negative feedback* reins things in: crowding triggers dispersal; depletion reduces exploitation; overheating increases ventilation. The art of swarm design, in nature or in algorithms, is to couple these two so that a system can both commit and remain sane.

A good way to feel this is to begin with something that looks like almost nothing: a trace.

In the late 1950s and 1960s, the French biologist Pierre-Paul Grassé was trying to describe how termites build such elaborate nests without blueprints or foremen. His answer was a concept he called *stigmergy*: coordination through the environment. The idea is beautifully spare. One individual changes the world—drops a pellet, lays a chemical mark, digs a small depression—and that change becomes a signal that biases what the next individual does. No termite needs to see the final cathedral. It only needs to respond to a local cue that earlier work has made more likely.

A trace is a kind of external memory. It is memory that does not require a brain to store it, and it does not have to be carried by the same individual who created it. This is what makes stigmergy so potent. The environment becomes a shared notebook, and the swarm writes in it with actions.

There are two broad ways a trace can work, and the distinction matters. Sometimes later actions simply strengthen the same signal. An ant trail is the canonical example: each passer-by adds pheromone, making the trail progressively more likely to be followed. That is *quantitative stigmergy*, because the trace changes in degree. Sometimes a trace changes what kind of action is triggered. A small pile of building material might cue a termite to add more; once the pile reaches a certain shape, it might cue excavation nearby rather than deposition. That is *qualitative stigmergy*, because the trace changes in kind. Both allow simple creatures to produce structures that feel planned, because the plan is stored in the world as a sequence of nudges.

If this begins to sound like human life, it should. A footpath across a lawn is stigmergy in sneakers: the more people cut the corner, the more the grass thins, the clearer the route becomes, and the more the next person is tempted to follow it. No committee meeting is required. The environment records preference and turns it into momentum.

Once you see stigmergy, the "alchemy" becomes less mysterious. The swarm is not conjuring order from nothing; it is building order by repeatedly feeding its own small outputs back into its inputs. A trace makes a behavior slightly more likely; the behavior reinforces the trace; the reinforcement makes the behavior much more likely. A tiny asymmetry becomes a highway.

The most important word in that description is *slightly*. Swarms often do not need a strong signal at the start. A weak bias is enough, provided the system can amplify it. That is why accidents matter in swarm systems. The first ant to discover food in one direction rather than another can set the colony's economy for the next hour. The first few birds to tilt left can pull a flock's whole arc. The first eddy in a stream can determine where foam collects. When the rules contain amplification, history leaves fingerprints.

This is where emergence stops being a slogan and becomes a way of thinking. When a system has positive feedback, you should expect it to be sensitive to initial conditions, to display sudden commitments, and to show path dependence—the tendency for the past to keep shaping the present even when circumstances change.

But real swarms do not merely amplify. If they did, they would be brittle, locked into whatever happened first, careening into runaway extremes. A functioning swarm needs brakes as well as engines, and those brakes are often local and automatic.

Ant trails, for example, are not just reinforced; they also decay. Pheromones evaporate. Rain washes them away. The chemical is a clock. If food disappears, the trail stops being refreshed, and the colony gradually forgets it. That forgetting is not a failure; it is a built-in negative feedback that prevents the colony from clinging to yesterday's opportunities. The colony can look decisive because it commits strongly when reinforcement is high, and it can look flexible because the memory it relies on is intentionally short-lived.

This pairing of amplification and decay shows up all over collective behavior. A fish school can tighten when threatened, but it cannot tighten indefinitely, because collisions become too costly and individuals peel away. A bee swarm can accumulate support for a potential nest site, but the support does not simply grow without bound; it competes with other sites, and there are costs to recruiting too aggressively. Even in purely physical systems, similar patterns appear: a sand pile grows steep until small avalanches relieve it; a traffic wave builds until drivers brake, dissipating the pulse.

The phrase "simple rules" can be misleading here. The rules are often simple for the individual, but their interaction creates a landscape of possibilities. In models of flocking, for instance, a tiny adjustment in how strongly an agent aligns with its neighbors can shift a group from milling to coherent travel. In social insects, a small change in the probability of switching tasks can reorganize labor allocation. The individual rule may be no more complex than "if you bump into enough nestmates quickly, keep doing what you're doing," but the collective can show thresholds that feel like decisions.

Thresholds are crucial because they let a swarm behave as if it had a shared standard without any shared deliberation. A threshold rule says: respond only when a signal is strong enough. Below the threshold, nothing happens; above it, action begins. The threshold can live in a nervous system, or it can live in the environment. A pheromone trail may need to reach a certain concentration before it becomes attractive. A pile of pellets may need to reach a certain shape before it becomes a building site. A quorum of bees may need to reach a certain count before the swarm commits.

That last example is worth lingering on, because honey bee swarms provide one of the most vivid cases of emergence producing something that looks like collective judgment.

When a honey bee colony becomes too large, it can split: the old queen leaves with a swarm of workers and clusters nearby, forming a temporary bivouac. While the swarm hangs, scout bees search for potential nest cavities. Each scout returns and performs a waggle dance whose vigor reflects the quality of the site she found. Other scouts may be recruited to inspect the advertised location, and if they agree, they dance too. Competing sites compete through competing recruitment. A swirl of dancing on the swarm's surface becomes, in effect, an information market.

No bee sees the full set of options. No bee tallies the votes. Yet the swarm converges on a single site with impressive reliability. One reason is the presence of a quorum mechanism: once enough scouts are simultaneously present at one site, their behavior changes, and the swarm's internal dynamics tip toward departure. Unanimity is not required; what matters is crossing a locally sensed threshold that signals, indirectly, that the option has enough support to be worth committing to. The swarm then lifts off in a coordinated way, as if an announcement had been made.

The same logic can be seen in a far less glamorous insect: harvester ants in the desert. These ants face a daily choice that is easy to state and hard to solve: how many foragers should the colony send out? Send too few and you miss food. Send too many and you waste energy or expose ants to deadly heat. There is no ant with a weather forecast and a colony-wide dashboard. And yet colonies adjust their foraging intensity minute by minute.

One of the mechanisms studied in harvester ants relies on local interaction rates: brief antennal contacts in the nest entrance area. Returning foragers, carrying food or empty-handed, bump into potential outgoing foragers. The rate of these contacts provides a fast, local proxy for current conditions outside. If returning foragers arrive frequently, outgoing activity increases; if returns slow, outgoing activity declines. It is a feedback controller built out of collisions and probabilities. The "signal" is not a message like "conditions are good"; it is a measurable flow of bodies.

This is emergence at its most practical. The colony does not need to estimate the environment explicitly. It needs a variable that tracks the environment well enough to regulate behavior. Local interaction rate becomes that variable. The whole system is tuned so that a few seconds of experience at the entrance can shape colony-level allocation, without any ant needing a model of the day.

Such examples can be told as charming natural history, but they also have a more unsettling implication: swarms can implement algorithms.

An algorithm, in this context, is not a line of code; it is a rule that maps inputs to outputs. "If the pheromone is stronger, follow more confidently." "If you meet many returning foragers, go out." "If the number of scouts at a site exceeds a threshold, switch from recruiting to preparing for departure." These are algorithms distributed across bodies and the environment.

The reason the results can look intelligent is that many problems of collective life are, at root, problems of *search* and *allocation*. Where is the food? How should effort be divided? Which option should be chosen? How should risk be managed? Humans solve these with explicit communication and central planning when we can afford it. Swarms solve them with local signals that have been shaped by evolution to track the relevant variables, and with feedback architectures that amplify good opportunities while damping runaway errors.

The most persuasive demonstration of this logic came, oddly, from a time when biologists had little interest in insect minds and mathematicians were thinking about cold atoms and magnets. In statistical physics, there is a long tradition of explaining large-scale order by local interactions: spins that align with neighbors can create magnetism; molecules that attract at short distances can create phase separation. When researchers like Vicsek and collaborators brought similar thinking to moving agents, they showed that collective motion could arise as a robust macroscopic state. In that framework, the flock is not a miracle; it is a phase.

Yet the physical analogy can mislead too. Molecules are indifferent; animals are not. Animals have goals, however small: to avoid collision, to keep up, to find food, to stay alive. The local rules are shaped not just by mechanics but by selection. That means the rules can exploit the world's structure in ways that a generic physics model cannot anticipate. An ant's pheromone decays at a useful timescale; a bee's dance encodes direction and distance; a fish's sensory system filters for the frequencies created by nearby tailbeats. Emergence in biology is not only about what can happen; it is about what works.

One of the most interesting things about biological swarms is how they make use of *constraints*. Constraints are often treated as limitations, but in collective systems they are a kind of scaffolding. If every ant could go anywhere at any time with equal ease, trails would be harder to form. If bees had no cost to dancing, recruitment would become noisy propaganda. If fish could occupy the same space without collision, the pressure to maintain spacing would vanish and the school would lose a stabilizing geometry.

Constraints also include the simple fact that individuals are embodied: they take time to turn, they have a maximum speed, they can only carry so much. Those limits can create regularities. A trail's width reflects how many ants can pass. A milling pattern reflects turning radius. A building's shape reflects what can be carried and where it sticks. The swarm's architecture is as much a product of bodies and materials as of rules.

This brings us back to the sensation of alchemy: something valuable created from ingredients that appear too plain. The trick is that the value does not come from any one ingredient. It comes from the loop. Local action changes the local world; the changed world changes the probability of future actions; the changed probabilities change the collective pattern; the collective pattern reshapes what is locally experienced. The loop closes, and once it closes, it can sustain structures far larger than any participant's awareness.

A useful way to test whether emergence is doing the work is to ask what happens when you remove the supposed "clever" element. Remove the idea of a map from the ant: trails still form. Remove the idea of a leader from the flock: coordinated motion still appears. Remove the idea of a central vote counter from the bee swarm: a nest site can still be chosen. The intelligence is not absent; it is relocated. It lives in the architecture of interactions.

There is a final twist that makes swarm emergence feel almost philosophical. Because the rules are local, the group pattern is not a goal that individuals are trying to achieve. It is a side effect that becomes useful. A trail is not drawn because ants want a trail; a trail emerges because following a stronger scent is a good local heuristic, and the resulting trail then makes foraging more efficient. A quorum is not counted because bees want democracy; a quorum emerges because waiting for a certain density of agreement reduces the risk of premature commitment, and the resulting threshold makes collective action possible.

Once a pattern exists, it can begin to shape the evolutionary future. Colonies that happen to have pheromones with the right decay rate exploit food better; those colonies thrive; the trait spreads. Swarms that happen to couple recruitment with inhibition avoid deadlocks; those swarms persist. Over time, evolution becomes a craftsman of feedback loops, tuning parameters that no individual can name, but that determine whether the crowd becomes a traffic jam or a flowing highway.

From a distance, the result still looks like a conductor. Up close, it looks like countless tiny choices, each made without overview, each responding to a cue that exists only because others have been responding too. The alchemy is not that simple creatures become complex thinkers. The alchemy is that simplicity, repeated and coupled, can produce patterns that are richer than any participant could imagine, and reliable enough to bet a life on.

1.3 The Queen's Illusion

Stand near an observation hive and it is hard not to tell a story with a throne in it. Thousands of honey bees stream across comb like a living fabric. In the middle of the press, somewhere you cannot easily see, is a queen: longer-bodied, surrounded by attendants, the mother of almost everyone in the room. The name does the rest. "Queen" is not a neutral label. It arrives with a suitcase of assumptions about command, obedience, and hierarchy. If a human city had a single indispensable individual at its reproductive center, we would not hesitate to call her a ruler.

And yet the closer you look, the stranger the metaphor becomes. A honey bee queen does not patrol. She does not inspect work. She does not bark chemical orders like a general. Much of her day is spent doing something intensely unglamorous: laying eggs.

The colony's spectacular coordination—foraging, nursing, ventilation, defense, comb construction—is carried out by workers who are not waiting for instructions so much as responding to one another and to the state of their shared environment.

The "queen-as-conductor" story is a particularly persistent illusion because it flatters a human habit of explanation. We feel safer when we can point to a single cause. We like to locate agency in a face. When something complicated works well, we look for the person in charge. When something fails, we look for someone to blame. It is psychologically easier to imagine a monarch than to imagine a distributed control system, because distribution has no obvious handle.

The irony is that distributed control is not exotic. It is familiar from many corners of human life, once you stop insisting on a leader. A market sets prices without a price czar. A traffic jam forms without a traffic jam manager. Languages evolve without a grammar committee that can actually keep up. In each case, individual decisions coupled through shared constraints create a collective pattern. Swarm systems simply lean harder into this logic because individuals rarely have the information or the bandwidth for centralized governance.

The temptation to anthropomorphize is strongest in eusocial insects—ants, bees, termites—because their societies look uncannily like ours. They have a division of labour. They build. They farm. They wage war. They maintain climate control. They even have something that looks like a class system: reproductive queens and sterile workers. If any creature invites political metaphors, it is an ant.

Those metaphors have a long history. In the nineteenth century, when natural history was often written as moral instruction, ant colonies were invoked as lessons in industriousness and obedience. "The ant" in Proverbs is a model labourer. Victorian writers described the "queen" as a sovereign, workers as loyal subjects, soldiers as patriots. Later, when human societies became preoccupied with bureaucracy and efficiency, insects were recast as tiny factory hands. The metaphors changed with the times, but the basic move remained: explain order by projecting a familiar hierarchy onto an unfamiliar system.

The cost of that move is that it blinds you to what the insects are actually doing.

A queen certainly matters. Remove her and the colony's long-term future collapses. Introduce a new queen and the colony's developmental trajectory changes. In that sense, she is an essential component. But essential components are not always commanders. Your heart is essential, and yet it does not "lead" your liver; it participates in a web of signals that keeps the organism viable. The queen is closer to a high-impact organ than a monarch. She sits inside a regulatory network whose other nodes are workers, brood, nest architecture, and environmental conditions.

The queen's influence is often discussed in terms of pheromones, which can sound like chemical diktats. Pheromones are chemical signals that carry information between individuals of the same species. In honey bees, the best-known suite is called queen mandibular pheromone (QMP), produced primarily by glands near the queen's jaws. It affects worker physiology and behavior in many ways, including retinue behavior (workers clustering around and feeding the queen) and aspects of reproductive suppression.

Popular accounts sometimes treat QMP as if it were a single "control substance"—a chemical crown. But biology rarely works by single points of failure. The effects associated with queens typically arise from a blend of signals, including other glandular products and cues that are not pheromones in the narrow sense at all. A particularly telling observation is that QMP by itself is not sufficient to reproduce the full suite of queen effects in every context, and queens lacking major mandibular components can still exert colony-level influence. Rather than a master command, the queen seems to be embedded in a redundant signaling system: multiple channels, overlapping effects, robustness against noise and damage.

That redundancy is exactly what you should expect if the colony is regulated by distributed feedback. In a distributed system, "control" is not an arrow from ruler to subject. It is a loop. Workers respond to queen cues, but queen cues are also shaped by the queen's physiological state, which is shaped by worker feeding and care, which is shaped by brood demand and food influx, which is shaped by foraging conditions, which is shaped by weather. The queen is not outside the system issuing orders; she is inside it, both influencing and being influenced.

A good way to see the difference is to think in terms of what engineers call a controller. In a thermostat-controlled room, the thermostat does not command the furnace out of authority; it measures a local variable (temperature) and triggers actions when thresholds are crossed. The "intelligence" is in the feedback loop that stabilizes a value. In insect societies, many such loops operate at once, often without any specialized measuring device. The measurements are embodied in encounter rates, odor concentrations, hunger, temperature, and mechanical vibration. The triggers are changes in probability: a worker is more likely to forage, more likely to fan her wings, more likely to feed larvae, more likely to attack an intruder.

Within that picture, queen signals begin to look less like decrees and more like inputs into a control circuit. They are informative, not authoritarian. A strong fertility signal can tell workers, in a chemically honest way, that the colony has a healthy reproductive engine and that workers should invest in worker tasks rather than activating their own ovaries. But this interpretation is not uncontested, and the debate itself is revealing. Some researchers have argued that queen pheromones could function as manipulative control mechanisms that suppress worker reproduction for the queen's benefit. Others argue they are honest signals of fertility or quality that workers respond to because it serves the colony's inclusive fitness. Experiments in ants such as *Temnothorax* have been used to probe whether queen pheromones align with fertility and whether worker responses look more like enforced suppression or conditional agreement.

The important point, for understanding swarm minds, is not which side wins in every species. It is that even if queen signals have coercive effects, they do not create a centralized command structure in the way human political metaphors suggest. A pheromone does not specify who should forage at 10:17 a.m. and who should ventilate the brood at 10:18. It does not route traffic across the comb. It does not tell builders where to place the next wax cell. It biases the colony's internal state, shifting the probability landscape within which workers interact. The fine-grained coordination still arises from worker-to-worker coupling and environmental traces.

There is another reason the monarch metaphor persists: it confuses reproductive centrality with informational centrality. Queens are reproductive bottlenecks; many colonies depend on a single queen to produce the next generation. Humans are primed to equate bottlenecks with control because in our systems the person who controls resources often controls decisions. In biology, resources and decisions are not always aligned. The queen controls the flow of eggs, but most moment-to-moment decisions are implemented by workers responding to local cues.

A concrete example makes this easier to believe. Consider how a colony regulates its foraging. In a human organization, you might expect a manager to assign shifts: "Send 200 workers to the field."

In many ants, foraging intensity can be regulated by local interaction rates among workers, especially near the nest entrance. Brief antennal contacts between returning and outgoing workers provide a stream of information about current success and risk outside. If returns are frequent, outgoing foraging ramps up; if returns slow, it tapers. No queen needs to be consulted. A queen could not possibly track these minute-by-minute fluctuations even if she wanted to. The regulation lives in the worker network.

Once you are sensitized to this, the "queen" begins to resemble something like a chemical context setter rather than a planner. Her presence can suppress worker reproduction and stabilize colony identity; her fertility state can influence worker allocation over longer timescales; her loss can trigger emergency rearing of new queens. But the colony's adaptive agility is rooted in decentralized feedback among workers.

The same logic helps dissolve a related myth: the idea that termite mounds are designed like human buildings, with termite architects implementing a plan. Termite mounds are extraordinary structures, and it is natural to see them as products of foresight. Popular accounts sometimes claim they function as sophisticated air-conditioning systems, with carefully engineered chimneys and vents. Some mounds do show impressive regulation of internal conditions, but measurement-driven critiques have repeatedly warned against overconfident narratives that treat every mound as a deliberate climate machine. Different termite species build different architectures; environments differ; and a structure that looks like a ventilator to a human eye may be, in mechanistic terms, the byproduct of stigmergic building rules responding to humidity, CO_2, temperature gradients, and material properties. When a system is capable of producing complex forms from local rules, it becomes dangerously easy to retrofit intention onto the result.

This matters because the "invisible conductor" is a story we tell ourselves, not a thing we detect directly. If you begin with the assumption that there must be a leader, you will look for one and interpret ambiguous evidence in its favor. The queen's central location, the workers' attention to her, the colony's dependence on her fertility—all of these can be marshalled as proof of command, even when the actual mechanisms are distributed.

There is a simple test that often breaks the spell: ask what kind of information would be required for central command, and whether any individual plausibly has access to it. To direct a foraging force intelligently, you would need real-time knowledge of food availability, predation risk, weather, and the internal state of the colony's stores and brood. To direct construction, you would need information about spatial layout, material availability, and structural integrity. To direct defense, you would need rapid detection of threats at multiple points of entry. No single insect has sensors in all these places. The colony has them collectively, because workers are distributed across the relevant sites. The architecture of control follows the architecture of information: it must be distributed because information is distributed.

A second test is to look at robustness. Hierarchical systems often have fragile points: remove the hub and the spokes lose coordination. Many swarms do not behave this way. They can lose large fractions of workers and still function, albeit less efficiently. They can be split and still form coherent subgroups. That resilience is exactly what distributed control buys you. Even when a queen is essential for reproduction, the colony's operational intelligence does not collapse into chaos the moment her signals fluctuate. The system is built to tolerate noise because the world is noisy.

This is not to romanticize swarms as utopias of decentralization. Distributed systems have their own vulnerabilities. They can get stuck in local optima. They can amplify wrong cues. They can be slow to reverse a strong positive feedback. They can be exploited by parasites that mimic signals. A monarch is not the only way to fail.

What the queen metaphor obscures is something subtler and, in a way, more impressive. A colony is not an army obeying orders; it is a network of workers continually adjusting their probabilities of action based on local contacts and shared traces. The queen is a powerful node in that network, but she is not its mind. The mind is in the pattern of interactions.

One can watch this most vividly during the moments when a colony must switch regimes. When a queen dies, workers do not wait for an instruction to panic. They respond to the absence of certain cues, to changes in worker physiology, to the state of larvae that can be turned into queens under the right feeding. The colony moves into emergency mode not because a leader declares it, but because many local thresholds are crossed at once. The drama is real, but it is not theatrical.

It is hard to give up the monarch story because it feels like giving up explanation. If there is no leader, what is the cause of order? The answer is not that there is no cause. The cause is simply harder to point at. It is a web of feedback loops, tuned by evolution, operating through bodies and environments that store and transmit information in chemical gradients, encounter rates, vibrations, and heat.

Once you start noticing distributed control, it is difficult to stop. A queen is then no longer a tiny sovereign but a biological constraint and a signal source, essential yet limited, powerful yet embedded. The colony's coordination becomes less like obedience and more like consensus without conversation: thousands of small adjustments, each made for local reasons, braided together into a society that holds.

Chapter 2

Chemical Conversations

Individual ants possess limited memory, yet the colony remembers everything by offloading data into the environment itself. By investigating volatile pheromones, decaying trails, and frantic physical collisions, we uncover how swarms build a dynamic, externalized brain that maps the world in real-time.

2.1 Writing on the World

Walk through a city after a snowfall and you can read what happened without meeting a single witness. Someone hurried across the street, someone circled the park, someone stopped at the bakery and lingered. The scene is silent, but the ground has become a page: a record of intentions translated into marks. For many social insects, the world is not merely a place where communication happens; it *is* the communication medium. The shared environment becomes a notebook that everyone can consult, update, and overwrite, even if no individual understands the whole story.

The technical name for this is *stigmergy* (from Greek roots meaning something like "stimulation by work"). The idea is simple enough to explain at a dinner table: an individual changes the environment, and that change makes certain actions more likely for others. No meeting is required. No plan has to be explained. The coordination comes from a loop that runs through the world, not through a central mind.

There is a special satisfaction in this kind of communication because it looks like planning from the outside while being built from tiny, local decisions. A single ant is not a strategist. It has no map of the terrain, no overview of the colony's needs, no internal model of where everyone else is. Yet the colony can, over minutes and hours, carve out a network of reliable travel routes to food, defend those routes, and abandon them when they stop paying. Stigmergy is one of the mechanisms that makes this possible: a way of turning the environment into a shared memory and a shared control system.

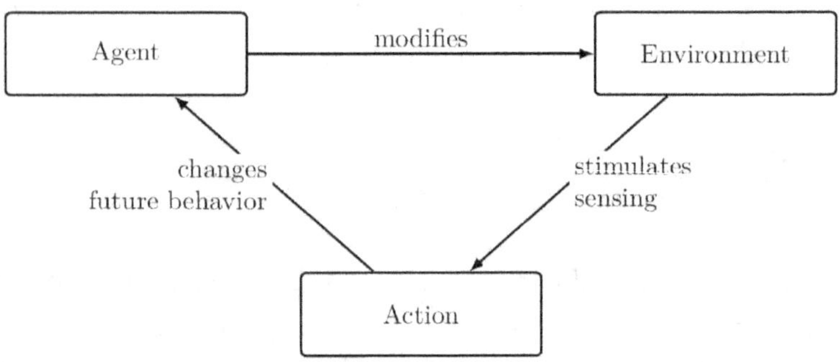

Stigmergy loop: work leaves traces; traces guide work.

The loop in the diagram is deliberately abstract, because stigmergy comes in many forms. Termites build arches and pillars not by carrying blueprints, but by responding to the shape and chemical texture of what is already there. Some wasps add material where the structure "feels" incomplete. Humans do this too, more often than we notice. A half-finished to-do list on a kitchen counter is a stigmergic device: it changes what you do next without needing a conversation. So is a well-trodden footpath that appears across a lawn: the path is not planned, but once it exists it recruits more walkers, deepening itself in a feedback loop. The world, once marked, begins to steer its inhabitants.

For ants, one of the most potent ways to write on the world is chemical. A forager can leave a trail by depositing a tiny amount of pheromone—a chemical signal produced by the body that influences the behavior of other members of the species. The key thing is not simply that the chemical attracts others. The more profound trick is that the trail is a *public, external memory*: it persists after the ant has gone, so it can guide individuals who never met the original writer. The environment stores information that would otherwise have to fit inside a brain the size of a poppy seed.

It is tempting to imagine an ant trail as a neat arrow pointing to treasure. Reality is messier, and more interesting. A trail is a probabilistic suggestion: *if you are here, doing this kind of task, in this context, then moving that way is more likely to pay off.* An individual ant follows gradients, compares alternatives at junctions, and integrates chemical information with what it senses directly—odors from the food, the feel of the ground, the presence of nestmates, and its own recent experience. What looks like an instruction from far away is, at ground level, a set of local nudges.

This is where stigmergy earns its keep. Direct communication requires an encounter. Two ants can touch antennae and exchange cues, but that is limited by who happens to meet whom. A chemical trail, by contrast, is asynchronous. It allows communication across time. It also allows communication across crowds: a single ant can write something useful for hundreds of others, not by broadcasting, but by altering a common surface.

The historical roots of this idea are charmingly practical. Long before anyone spoke about stigmergy, naturalists watched ants and tried to work out how they navigated. Early observers described ants "tracking" one another with uncanny precision and argued about whether they used sight, smell, or some mysterious sense of direction. As experimental biology matured, researchers began to test these hunches with the kind of simple interventions that feel like party tricks but reveal deep principles: wiping the ground, shifting a bridge, turning a paper strip, or swapping the path segments between two colonies. When a trail-bearing surface was rotated and the ants turned with it, the message was clear: something on the substrate mattered. A route could be detached from the landscape and carried like a ribbon.

Those experiments also broke a romantic misconception: trails are not always permanent, and they are not always singular. Ants use a chemically diverse toolkit. In some species a trail pheromone may be one dominant compound; in others it may be a blend. Different glands can contribute depending on the species and the context, and a trail can carry more than one kind of meaning. That richness matters because a colony does not just need to know *where* to go; it also needs to regulate *how much* effort to spend going there. A trail is not merely a signpost but part of a control system.

To see how writing on the world becomes a kind of collective computation, imagine the simplest foraging story. A colony has many potential directions it could explore. Scouts fan out. One of them finds food and returns to the nest, laying a trail as it goes. Now the ground between the nest and the food has been modified. Other ants leaving the nest are more likely to step onto the marked route. Those ants, if they reach the food and return, may reinforce the trail by depositing more pheromone. The trail grows stronger with use.

This is positive feedback made physical. In economics it is the dynamic behind "the rich get richer"; in social media it is the algorithmic boost that makes popular posts more popular. In ant foraging it is why a colony can, quickly and without a manager, focus its labor on a profitable source. A small advantage—a slightly shorter route, a slightly better patch of food—can be amplified by reinforcement until the colony behaves as if it had compared options and made a decision.

But it is crucial to keep the perspective right. The colony is not running a conscious vote. The trail does not contain a full argument for why this food is best. The information encoded in the environment is minimal and local: a chemical intensity along a path, perhaps with subtle variations at junctions. Yet that is enough. The algorithm is in the loop: write, read, respond, write again.

A concrete example makes the logic tangible. Consider a branching path, like a fork in a garden trail. Two routes lead to the same food. If one route is slightly shorter, an ant traveling it completes the round trip faster. That ant therefore has more opportunities per unit time to add pheromone to the path. The shorter route gets reinforced more frequently and becomes more attractive. Soon most ants are using it. The colony has effectively discovered a shortest path without any ant measuring distances. The computation is outsourced to physics (travel time) and chemistry (pheromone accumulation). This sort of stigmergic selection is one reason engineers became fascinated by ants and borrowed their principles for routing data through communication networks.

The beauty of the mechanism is also its vulnerability. Positive feedback can lock a group into a choice even when the world has changed. A trail, once strong, becomes self-justifying. Ants following it add more pheromone, which makes it stronger, which recruits more ants. That is why the concept of writing on the world must include an implicit editor: the world is not only a notebook, it is a notebook in which ink fades. Chemical trails are powerful partly because they have a built-in expiry date. Even before turning to evaporation and forgetting in detail, it helps to notice that persistence is a tuning parameter. A trail that lasts seconds is more like a shout. A trail that lasts hours is more like a signpost. The colony benefits from having both, and from having a way to regulate how long its messages remain legible.

Even at the "writing" stage, the act is not automatic. Ants do not deposit trail pheromone continuously like a leaky pen. They follow rules that depend on context: how rewarding the food is, whether they are laden, whether they are returning to the nest or leaving it, whether the path is crowded, and how confident they are. In the black garden ant *Lasius niger*, for example, trail laying interacts with the ant's own route experience: pheromone can reduce errors on complex routes, but experienced foragers also carry private information in memory. When private memory and public trail conflict, behavior can shift in ways that reveal negotiation rather than blind obedience. A trail is not a dictator; it is an argument the environment makes, and each ant decides how seriously to take it.

That interplay between public and private information is easy to recognize in human life. A restaurant with a line outside acts as a pheromone trail: it draws you in by broadcasting others' choices. But if you have been there before and know the food is mediocre, you may ignore the line. If you are starving and unfamiliar with the neighborhood, you may treat the line as strong evidence. Ants face the same kind of inference problem, scaled down to scent and touch.

There is also an important philosophical point hidden in this chemistry: stigmergy allows a society to have something like a shared memory without any individual being in charge of remembering. Memory, in this sense, is not a private hoard inside a skull; it is a distributed pattern maintained by many small acts. This makes the system robust. If a particular ant dies, the trail does not vanish. If many ants are distracted, the trail can persist for a while, carrying the information forward. The information is not located in a leader; it is smeared across the ground.

Of course, smearing across the ground comes with trade-offs. The environment is not a secure database. Rain can wash trails away. Sunlight and heat can alter chemicals. Wind can redistribute them. Other animals can trample the substrate. Rival colonies may exploit or interfere with markings. Even nestmates can create noise by leaving overlapping trails to multiple targets. The colony thrives by treating the world as a usable but imperfect medium, one that requires continual maintenance and continual interpretation.

Once you begin to see pheromone trails as writing, the world of an ant becomes strangely intimate. The ground is full of meaning: not just food trails, but boundary markings, alarm signals, nest odors, and the faint signatures of frequent passage. Some species lay persistent "footprint" cues simply by walking, leaving hydrocarbons that accumulate over time; others rely on more volatile compounds for rapid recruitment. The result is not a single message but a layered text with different half-lives. A busy entrance becomes saturated with the chemical equivalent of fingerprints. A fresh trail to a new food source is more like a wet line of ink. Both can guide behavior, but in different ways.

Stigmergy also reframes what it means to "coordinate." Coordination is often imagined as a kind of agreement reached through dialogue. Ants show a different possibility: coordination as alignment with a changing landscape of cues. When an ant follows a trail, it is not obeying a specific individual. It is responding to the accumulated work of many. In that sense, stigmergy is deeply democratic and deeply conservative at the same time. Democratic because every contributor can, in principle, add to the shared signal; conservative because existing marks bias future actions, so history has weight.

The weight of history can be a blessing. It allows a colony to exploit reliable resources efficiently. It also allows a colony to build stable infrastructure: trunk trails that become the equivalent of roads, especially in species that repeatedly harvest from the same areas. But the same weight can become inertia. If conditions shift quickly, a strong trail can keep ants pouring effort into yesterday's opportunity. The colony lives on the edge between persistence and flexibility, and chemical writing is one way it negotiates that edge.

There is a final twist that makes stigmergy feel less like a metaphor and more like a design principle. When an ant modifies the environment, it is not merely sending a message; it is changing the future state of the system in a way that shapes what signals will be produced next. A stronger trail recruits more ants, which increases traffic, which changes crowding, which can alter how ants deposit pheromone and how fast it is worn or replenished. The environment is not a passive noticeboard. It is an active part of the feedback loop, with its own dynamics.

That is why the simplest description of stigmergy is also the most useful: the work of others stimulates further work. A trail is work made legible. It is effort converted into information. It is a way for a society to extend its mind into the world, not by making individuals smarter, but by making the world more informative.

And once you start looking for this kind of externalized intelligence, it appears everywhere. A desire path across grass is a pheromone trail written in crushed stems. A comment thread that keeps resurfacing because each reply pushes it upward is a digital trace recruiting more attention. Even a stack of used plates at a busy café is stigmergic: it signals to staff where to clean first, without anyone issuing an instruction. Ants did not invent the trick, but they refined it with chemical ink and a talent for turning tiny acts into collective order.

2.2 The Essential Art of Forgetting

A trail is a promise written in a chemical alphabet: *food this way; effort will be rewarded.* The problem with promises is that the world changes faster than ink can dry. A patch of crumbs is swept away. A flower stops producing nectar. A rival colony finds the same prize. If the mark on the ground remained as persuasive tomorrow as it was today, a colony would spend its life chasing ghosts.

For a swarm, forgetting is not a failure of memory; it is a design feature. The colony does not merely need a way to store information in the environment. It needs a way to *delete* information that has become misleading, and to do so at roughly the same pace as the world updates. In ant societies, this deletion happens through two intertwined forces: the physics of chemicals in air and sunlight, and the behavior of ants deciding when *not* to reinforce a signal.

It helps to begin with a common misunderstanding. People often talk about pheromone trails as if they were permanent roads—once laid, forever available. In reality, many trail pheromones are volatile: they evaporate, oxidize, or break down. That fragility can sound like a limitation, as if ants were stuck with unreliable ink. But fragility is exactly what keeps the collective mind from becoming cluttered. A short-lived chemical trace acts like a timestamp: strong means recent; weak means old; absent means either never or no longer. The environment itself does the bookkeeping.

A human analogy is the difference between a live navigation app and a paper map. A paper map is stable and reassuring, but it does not warn you that an accident has closed the road. The value of the app is not only that it remembers; it also forgets quickly. Yesterday's traffic jam does not get to vote on today's route. Ant trails achieve something similar, not by connecting to satellites, but by letting chemistry fade at a biologically useful rate.

This need for forgetting becomes vivid when you consider what a colony faces each morning. Temperature rises. Humidity changes. Scents disperse differently. A trail that was informative at dawn may become noisy at noon. In some species, warm conditions can shorten the effective lifespan of a trail dramatically, not because the ants become less intelligent, but because the medium becomes less stable. Heat increases evaporation and can accelerate chemical decay; the consequence is that the same mark can carry less information simply because the air is warmer. On a blistering day, the colony is forced into a more present-tense style of communication: only the newest traces count for much.

That physical dependency ties collective intelligence to microclimate in a very direct way. When the substrate heats up, a trail can become harder to discriminate from the background; the chemical gradient flattens. A forager arriving at a fork is no longer reading crisp letters but smudged handwriting. If you have ever tried to follow a pencil note left in the rain, you know the feeling: you can still see that someone wrote something, but the words have lost their edges. In the same way, temperature can blur chemical information, and what looks like indecision in a line of ants may be a sensible response to reduced signal quality.

Yet physics alone does not explain the neatness of many trail systems. If evaporation were the only force at work, trails might still persist long enough to cause trouble, especially around heavily used routes. The more a trail is used, the more it is reinforced; and the more it is reinforced, the more it recruits. Without a counterweight, positive feedback becomes a kind of addiction: the colony keeps returning because it keeps returning. The result would be not efficient exploitation but runaway crowding, a jammed highway to a depleted site.

Negative feedback provides the counterweight, and it comes in forms that are surprisingly subtle. One possibility would be an explicit *stop* signal: an inhibitory pheromone laid to repel nestmates from a trail. Some insects do use inhibitory signals in certain contexts, but ants often do something more economical. They simply reduce how much they add to the trail when conditions suggest the signal is already too strong, too crowded, or no longer worth escalating. In other words, the colony edits by withholding ink.

Crowding is a particularly elegant trigger for this restraint because it converts a problem into information. A congested path is not just inconvenient; it is evidence that traffic is already high. If every ant reacted to congestion by depositing *more* pheromone—as if shouting louder in a noisy room—the trail would become even more attractive, drawing yet more ants into the bottleneck. What prevents this spiral is that crowding can suppress deposition: fewer ants lay pheromone on crowded paths, so the trail stops strengthening, and alternatives become comparatively more appealing. The colony is, in effect, using its own discomfort as feedback.

This is the same logic that makes human crowds self-regulating when they work well. A packed restaurant can attract customers up to a point, but if the line becomes intolerable, people stop joining it. The restaurant loses its ability to recruit more diners through popularity alone. Ants implement a comparable policy without anyone deciding it explicitly. Each forager responds locally to the density of encounters and the difficulty of movement. The collective outcome is a throttling of recruitment.

Forgetting, then, is not always the erasure of a trail that already exists. Often it is a refusal to keep re-writing it. The trail fades because the colony stops agreeing to pay the cost of maintaining it.

That cost is small per ant but large in the aggregate. Pheromone compounds are produced by glands, stored, and dispensed in minute quantities, but those quantities add up when thousands of ants lay trails repeatedly. A colony that squandered its chemical resources would be like a household that leaves every light on in every room: the waste is not dramatic moment-to-moment, but it becomes ruinous. Editing signals is a form of metabolic thrift as much as a form of information hygiene.

Information hygiene is not merely aesthetic; it can decide survival. Imagine a landscape dotted with brief opportunities: a fallen piece of fruit that will be stripped quickly, a dead insect that will soon be discovered by scavengers, a patch of seeds that may be exposed only until the wind shifts. In such a world, the most dangerous error is not missing a resource once. The most dangerous error is sending labor to a resource that is gone. An ant is time and energy on legs. Every minute spent marching to nowhere is a minute not spent finding something real.

This is why decay can be understood as a built-in mechanism for *relevance*. A strong trail means not only that a resource exists; it also implies that nestmates have encountered it very recently and found it worth returning from. The strength of the signal is, indirectly, a measure of ongoing success. A weak trail is ambiguous: perhaps the resource has dwindled; perhaps the route has become risky; perhaps the colony has shifted attention. The ambiguity is useful because it prevents the colony from treating an old trace as a command. Weakness invites exploration.

The idea that forgetting is an adaptive tool has an intellectual history that runs through early cybernetics and the study of feedback systems. Engineers learned long ago that control requires both amplification and damping. A thermostat would be useless if it could only heat and never stop heating. A public-address system would be unbearable if it amplified sound without suppressing squeal and echo. Ant trails, too, sit between growth and decay. Reinforcement allows a good route to dominate; decay and restraint prevent domination from becoming tyranny.

A concrete example shows how closely this resembles an engineered system. Picture two routes from nest to food. Route A is slightly shorter but passes through a narrow gap, so it congests easily. Route B is longer but wide and smooth. Early on, route A may win: faster travel leads to faster reinforcement. But as traffic increases, congestion slows travel and increases encounters. If crowding suppresses further deposition on route A, its pheromone intensity stops climbing. Route B, though initially weaker, can then become competitive because it is less clogged and can be reinforced steadily. The colony does not need to *understand* that a bottleneck exists. It only needs individuals whose trail-laying behavior responds to local crowding. The balance between two routes emerges from the interaction between movement, encounter rates, and chemical persistence.

The most striking feature of this story is that the colony can adapt without any individual holding a model of the whole network. Each ant experiences only a handful of cues: how often it bumps into others, how hard it is to move, how strong the trail is beneath its feet, and what reward it just received. Out of these fragments, the group builds a pattern that looks like traffic management.

Forgetting also protects the colony against a different kind of failure: information overload. When we imagine information overload, we usually think of human screens crowded with notifications. Ants face something like the same issue on the ground. The environment is full of smells, including the colony's own chemical background. If pheromone marks accumulated without limit, every surface would become saturated, and the contrast between meaningful trails and irrelevant residue would shrink. In signal-processing terms, the signal-to-noise ratio would collapse.

Decay preserves contrast. It keeps chemical landscapes readable by preventing yesterday's ink from tinting everything. This is another way in which forgetting is not the opposite of intelligence but one of its prerequisites. A brain that remembered every sensation with equal strength would be paralyzed by detail. Selective forgetting is what makes attention possible. Ant societies achieve an analogue of attention by allowing irrelevant traces to fade below threshold, so that only actively maintained routes stand out.

The word *threshold* is important here. Ants do not read pheromone concentration as a linear dial. Their behavior often switches when a cue crosses a certain point: follow versus explore, recruit versus refrain, turn left versus turn right. Thresholds turn continuous chemistry into discrete decisions. But thresholds also create danger: if a trail remains above threshold long after it is useful, it will keep triggering the same decisions. Forgetting is what pushes obsolete cues back below threshold, restoring flexibility.

Temperature again plays a role in setting these thresholds in practice, because it changes how quickly concentration falls. A trail laid under cool conditions may stay above threshold longer; under hot conditions, it may dip below threshold quickly, forcing ants to rely more on immediate cues or on their own route memory. The colony's "attention span" is partly a property of weather.

One might ask why ants do not solve all of this with better internal memory. Why not have each forager remember the route to food and ignore chemical clutter? In some cases, they do. Many ants can learn routes, use landmarks, and return efficiently without trails. But private memory does not scale as smoothly as environmental writing. A personal memory helps an individual; a trail helps anyone who happens to arrive. A personal memory dies with the ant; a trail can outlive her. Private information is also expensive: it takes neural tissue, development, and time to learn. A colony can offload much of that burden into the environment using tiny quantities of chemicals and simple rules of thumb.

Forgetting, then, is part of what makes this offloading safe. If the world is to serve as a memory store, it must also serve as a garbage collector.

It is worth lingering on the emotional resonance of this. We like to think of memory as precious, and forgetting as loss. In ant societies, forgetting is closer to mercy. It is the mechanism by which the colony frees itself from past commitments. A trail that fades is the colony releasing a plan that once made sense. There is something almost humane about a system that can let go.

Even more intriguing is that forgetting can be tuned, not only by climate, but by behavior. Ants can adjust how much pheromone they lay depending on how good the food is. Richer food tends to provoke stronger recruitment; weaker food provokes less. That means trail strength reflects not only where food is, but how valuable it has recently been. If the food quality declines, ants reduce reinforcement, and the trail begins to fade faster than it would under constant deposition. Value becomes legible in the landscape as a changing intensity.

This gives the colony a crude but effective economic sense: invest where returns remain high; divest where returns fall. The divestment is forgetting.

There is also a darker side, which makes the system feel less like magic and more like life. Forgetting can cause missed opportunities. A trail that evaporates too quickly may prevent efficient exploitation, forcing ants to rediscover the same resource repeatedly. The colony lives with this trade-off, and different species occupy different points along it, shaped by ecology. In stable environments with persistent resources, longer-lasting cues can be advantageous. In volatile environments, quick decay protects against wasted effort. No single rate is best in all worlds.

The most honest way to describe the collective intelligence here is not that it always finds the optimal solution, but that it remains *adaptable*. Adaptability requires openness to new information, and openness requires clearing space for it. A surface that holds every message forever becomes unreadable; a system that honors every past signal becomes rigid. Ant colonies avoid rigidity by letting chemistry return, slowly and continually, to silence.

In the end, the essential art of forgetting is an agreement between biology and physics. The ants provide reinforcement when it is useful and restraint when it is not. The environment provides decay, erasure, and the steady insistence that messages must be renewed to remain persuasive. Together they create a living text that stays current precisely because it is always disappearing.

2.3 Bumps, Tremors, and Touch

Chemical trails can make the ground behave like a notebook, but not every message can wait to be written, dried, and read back later. Sometimes the world is too noisy for chemistry to carry the day. Rain blurs odor. Wind shears a plume into tatters. Heat accelerates evaporation until yesterday's map becomes a faint rumor. And sometimes the colony is working in places where *time* is the scarce resource: a food bonanza that will be gone in minutes, a traffic jam at the nest entrance, a predator's shadow that demands an immediate, coordinated response.

In those moments, ants and other social insects lean on communication channels that are almost embarrassingly direct. They bump. They touch. They vibrate the ground. The signals are local, bodily, and fast. To human ears this can sound like a downgrade from the elegance of chemical writing. In practice it is the same principle expressed through a different medium: simple cues, read locally, that scale into collective regulation.

Touch is the most basic. An ant's antennae are not mere feelers; they are mobile laboratories. With a brief antennal tap, an ant can sample the waxy hydrocarbons coating another ant's body, which carry information about nest membership and, in some species, something about recent activity. But the most useful part of touch for a swarm is not the chemical detail it can convey. It is the fact that touch events are *countable*. In the dark interior of a nest, where vision is limited and air currents can scramble odor, the simplest question an ant can ask is: how often am I meeting others, right here, right now?

That question turns social life into a sensor. Encounter rate becomes a proxy for local density and local flow. If you are standing in a train station and people keep brushing past your shoulders, you do not need a CCTV camera to know the platform is crowded. Your body already knows. Ants can use the same logic, but in a tighter feedback loop: the rate of collisions can change behavior within seconds.

The power of this idea becomes vivid in a desert, where foraging is dangerous and timing matters. Red harvester ants (*Pogonomyrmex barbatus*) live in places where heat can cook an ant in minutes, and where food is scattered seeds rather than steady streams of nectar. A colony must decide each day how many foragers to send out. Too few, and it misses opportunities. Too many, and it wastes water and energy, and risks catastrophic losses to heat or predators. There is no safe way for a single ant to survey the whole landscape and make an executive decision.

Deborah Gordon's long-running work in Arizona made the solution famous: regulation without centralized control, built from brief interactions at the nest entrance. Inside the nest there is a kind of lobby, an entrance chamber where returning foragers pass outgoing workers. Those meetings are fast, often just a tap of antennae and a step around one another. Yet the *rate* of such encounters can act like a control signal. When returning foragers are coming back frequently, outgoing activity ramps up; when returns slow, departures slow too. The colony is not consulting a map of the environment; it is sampling the environment indirectly through the success rate of ants that already took the risk.

This is an important shift in what counts as "communication". A return rate is not a message deliberately sent in the way a trail is laid. It is a consequence of outside conditions: the availability of seeds, the temperature, the wind, the threat level. The nest entrance becomes a place where those outside conditions are translated into local interactions. A forager coming back is, in effect, a moving data point. The outgoing ant does not need to know *why* returns are high. It only needs to treat high return rates as evidence that foraging is currently paying.

What makes the system especially compelling is that it is fast. Chemical trails can guide traffic across meters or tens of meters. Encounter rates can update decisions on the timescale of seconds, which is closer to the timescale at which danger unfolds. A sudden cloud cover can make it safe to send more foragers; a burst of heat can make it suicidal. In a world where the difference between safe and fatal may be a few degrees, slow governance is bad governance.

Researchers tested the causal role of encounter rates with clever interventions that feel like stage magic but are really good experimental design. If the colony is using returning ants as a signal, then artificially increasing "returns" should stimulate outgoing foraging. One way to do this is with bead mimics: small objects coated with chemicals that smell like nestmates. Rolled into the entrance at controlled rates, these mimics can raise the encounter rate experienced by ants inside. When the encounter rate rises, the onset of foraging can shift earlier. In other words, the colony is not simply responding to time of day; it is responding to local interaction statistics.

This turns bumping into an information channel, and it also turns bumping into a safety mechanism. A traffic controller who only knows "more cars are coming back" is forced into caution when returns are low. That restraint is not a moral virtue; it is a collective risk policy implemented through a simple rule. Outgoing ants can treat low encounter rates as a warning that conditions outside are poor. They do not have to see the scorching ground to behave as if it is scorching.

There is a deeper lesson here about what a swarm chooses to measure. When chemical trails dominate, the colony measures *space*: where to go. When bumping dominates, the colony measures *rates*: how much to invest, how urgently to act. Both kinds of measurement can be implemented with tiny brains, because neither demands a global picture.

Collision-based sensing has a second virtue: it can work even when individuals are nearly identical. An ant does not need to recognize individuals; it only needs to register contacts. That makes the method robust to mistakes. If an ant misidentifies a particular nestmate, the error barely matters. If it misestimates encounter rate persistently, *then* the system shifts. Swarm intelligence often looks like clever cognition, but much of it is clever *statistics*: decisions guided by aggregated experience rather than by precise identification.

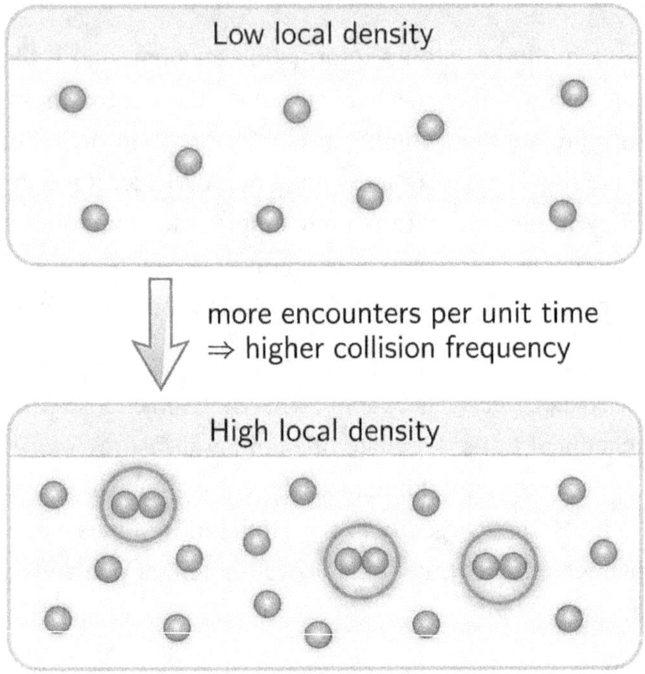

An individual can estimate crowding by counting contacts, without seeing the whole group.

Vibration adds another layer of immediacy, and it is easy to underestimate because humans are so visually biased. For an ant on the ground, the world is a drumhead. Footsteps, chewing, digging, and even the movement of wind through vegetation can set the substrate trembling. Some ants create vibrations deliberately by stridulation: rubbing body parts together to produce a patterned signal that travels through the ground or through plant tissue.

Leafcutter ants provide a particularly vivid example, because their lives are built around moving heavy cargo through dark, crowded spaces. A forager cutting a leaf fragment is engaged in strenuous work, and work creates disturbance. When leafcutter ants stridulate, the vibrations can function in multiple contexts: recruiting help, signalling alarm, coordinating excavation, and modulating traffic. The medium is well suited to places where air-borne chemicals may be diluted or delayed, such as on a windy canopy or within a dense, busy trail.

Vibration also allows something chemistry often cannot: directional guidance without a persistent track. A trail pheromone can create a spatial gradient, but reading it requires being on the substrate where the pheromone sits, and it may take time to build. A vibration source, by contrast, can be located through differences in arrival time and intensity across the legs. The ant's body becomes a kind of multi-sensor array. The result can resemble tropotaxis: comparing signals on the left and right sides to orient toward the source.

The elegance here is that the ant does not need to interpret the vibration as a complex "message" with grammar. It only needs to treat it as a beacon: *something important is happening over there*. In engineering terms, vibration can be a low-bandwidth but low-latency channel. That is often exactly what a swarm needs.

Of course, the world vibrates for many reasons, and that raises the problem of noise. Wind shakes plants. Raindrops strike leaves. Large animals browse. A useful system must avoid panicking at every tremor. Evidence suggests that some ants can discriminate ecologically meaningful patterns from irrelevant background, responding differently to, say, the heavy, rhythmic disturbances of a large herbivore compared with the more irregular flutter of wind. The colony's defensive behavior can then be coordinated by many individuals responding to a shared physical cue, without anyone having to spread a chemical alarm through a vast area.

The most interesting part of vibrational communication is not that it replaces chemical cues, but that it can interact with them. A colony does not live in one modality. An ant arriving at a busy trail junction may smell a faint recruitment trail, feel the jostle of bodies, and detect a distant tremor through its legs. These cues do not merely add up; they can gate one another. A weak chemical signal may be ignored when encounter rate is low, but acted on when traffic is high. A vibration that would normally trigger alarm may be discounted if the antennal contacts in the nest lobby suggest calm routine.

Humans do this constantly. A distant bang means one thing in a quiet neighborhood at midnight and another thing on a city street at noon. Context changes interpretation. In insects, context can be implemented through very simple mechanisms: hormonal state, recent experience, or the immediate density of neighbors.

There is a subtle emotional payoff in recognizing this. The intelligence of a swarm does not reside in one brilliant channel of communication. It resides in the ability to keep functioning when any particular channel becomes unreliable. A chemical trail is a superb external memory, but it can be wiped clean. Touch is immediate, but it is local. Vibration travels quickly, but it can be noisy. By mixing these channels, a colony can remain coordinated in darkness, heat, wind, and chaos, using the world itself as both messenger and test.

The result is a society that stays sensitive without becoming fragile. A forager does not need certainty; it needs enough evidence to take the next sensible step. Sometimes that evidence is a fading chemical line. Sometimes it is the simple fact that, in the entrance chamber, no one is coming home. Sometimes it is a tremor that tells the ground has begun to move with someone else's appetite.

Chapter 3

The Hive's Democracy

When a honeybee swarm splits, thousands of individuals must agree on a single new tree hollow within days or die of exposure. Without a president or committee, they solve this life-or-death problem through a rigorous competition of enthusiastic dancing and independent verification. This process reveals the stark mathematics of how nature achieves consensus without command.

3.1 The Dance Floor Debate

A honeybee swarm on a spring day can look like an emergency paused mid-action: a dense, living mass draped from a branch, humming in place while the world keeps moving. The old queen is there, somewhere inside the cluster, but she is not issuing orders. The swarm has a problem with a deadline. Exposed on a branch, the bees are vulnerable to rain, chill, predators, and simple exhaustion. Yet they do not blunder into the first dark hole they find. They deliberate.

The strange part is where that deliberation happens. Not in the air, not at the candidate home, but on the surface of the swarm itself. The clustered bees become a kind of public square. Scouts land, squeeze through the outer layer, and climb upward until they reach open space and light. Then, with the purposeful intensity of someone who has found a vacant apartment in a tight rental market, they begin to advertise.

A scout's advertisement is the waggle dance: a figure-eight path with a straight central run where the bee waggles her abdomen while vibrating her wings. In a foraging context, the dance encodes direction and distance to flowers; in house-hunting, it does the same for potential nest sites. But for the swarm, the dance carries another message that is less like a map and more like a review score. How *much* a bee dances matters. The number of circuits she performs, and how persistently she repeats them over minutes, becomes a measure of her enthusiasm. Many bees can be wrong in the details, but when you let them compete openly, stronger proposals tend to recruit more attention.

Imagine the swarm's surface as a dance floor with limited space and limited audience. A handful of dancers can easily be ignored. A dancer surrounded by attendants becomes difficult to miss. The difference is not charisma; it is mechanics. Following a waggle dance requires close attention: bees press their antennae to the dancer's body, tracking her waggling run, then looping around to watch again. Those followers are not passive. Some of them become recruits: they take off and fly to the advertised location, often with the dancer's coordinates as their initial guidance. The dance is thus both speech and replication: the message makes more messengers.

That creates the swarm's first and most important feedback loop. A scout who returns from a superb cavity tends to dance longer. A longer dance tends to recruit more followers. More followers means more scouts visiting that same site and potentially returning to dance for it as well. A mediocre cavity can still be proposed, but its advocate usually offers a shorter dance; fewer recruits investigate; fewer return impressed; the proposal withers.

This is not metaphorical democracy, the way we sometimes borrow political language to make animal behavior sound dignified. It is a literal, physical marketplace of attention built out of bodies. Each dance spends the dancer's time and energy. Each follower spends attention that could have been given elsewhere. Each recruit spends flight fuel and risk. Nothing is free, and those costs are precisely what keep the swarm from being swamped by noise.

The hard part is believing that a bee can evaluate a home well enough for the dance to carry meaningful information. A tree cavity is not simply "big" or "small." A good nest site for honeybees tends to have enough volume for comb storage, a small defensible entrance, and dryness and shelter. The scout does not carry a tape measure. She explores by walking the interior, tasting surfaces, sensing airflow, and gauging light and damp. Her assessment will be imperfect, and two scouts can disagree. Yet, across many scouts, the swarm gets something like an average opinion—not by taking a vote at the end, but by letting advocacy itself be graded.

Thomas Seeley and colleagues, working in the late twentieth century and early twenty-first, made this process visible by giving swarms a problem they could measure. They set out nest boxes of different quality: some roomy, dry, and well-insulated; others cramped or overly exposed. When scouts returned, they did not dance in identical bursts. Better boxes reliably drew longer bouts of dancing: more circuits per scout, and more repeated performances across time. The signal is noisy at the level of individuals, but at the level that matters—the total dance activity accruing to each candidate site—quality emerges as a statistical bias. It is the difference between one person's unreliable restaurant review and a crowded dining room on a Saturday night.

The swarm does not require that scouts be honest in the moral sense. It requires only that their behavior be tied to their own assessment strongly enough that the collective signal is informative. The waggle dance achieves this in a beautifully economical way: the dancer is not trying to persuade with words; she is simply performing more when she is more impressed. Because dancing is costly, the cost itself discourages unlimited exaggeration. A bee can "lie" only by spending her own limited time dancing for a site she does not actually rate highly. Evolution has no reason to favor that.

Yet even with honest signaling, why does the swarm not get trapped by the first site found? In many collective systems, early momentum becomes destiny. A rumor spreads because it spreads, not because it is true. If one scout happens to stumble on a mediocre cavity quickly, she can begin recruiting before any better option is even on the table. Momentum would then be a kind of accidental tyranny.

Honeybees counter this in a way that feels almost like procedural fairness. Early in the search, the first discoverers often dance even for sites that are not particularly good. In Seeley's work, this tendency is sometimes summarized as "discoverer-should-dance": the act of discovery itself triggers a burst of advertisement, as if the swarm rewards novelty with airtime. That bias makes sense. When little is known, the greatest danger is not choosing the second-best option; it is failing to sample the environment widely enough to find any good options at all. A brief, default dance for a newly found site seeds exploration. After that seed, the system becomes more selective. If the site is mediocre, the dance decays and recruitment dries up. If the site is excellent, the dance persists and recruits multiply.

The decay is crucial. A waggle dance is not a permanent billboard. It fades. Even for a high-quality site, an individual scout's enthusiasm wanes over time; she dances fewer circuits in subsequent bouts and eventually stops, unless renewed by revisiting the site or by the social energy of a growing campaign. For a low-quality site, the decay is faster. This built-in expiration prevents the dance floor from being clogged with yesterday's opinions. It is also how the swarm avoids endless stalemate when several sites compete: losing campaigns tend to die, not necessarily because their supporters are converted by argument, but because each advocate has a limited personal budget of dancing.

To picture the debate, it helps to picture the swarm's surface not as a tranquil stage but as a bustling newsroom. Scouts arrive at different times with different leads. Some are fresh from inspection and dance with vigor. Some are tired and stop. Followers weave among them, choosing whom to attend. Recruits launch, return, and sometimes join the chorus. The swarm is not comparing sites with a centralized list; it is letting proposals rise and fall through a distributed process of amplification and forgetting.

A concrete example makes the dynamics easier to feel. Consider a swarm hanging from a low apple branch at the edge of a meadow, with a line of old trees bordering the field and a farmhouse beyond. One scout discovers a hollow oak cavity: deep, dry, with a narrow entrance protected by a knot of bark. Another scout finds a gap under the farmhouse eaves: sheltered from rain, but with a broad opening and strong drafts. A third finds a birdhouse: charmingly enclosed, but too small.

All three return and dance. The oak scout dances for longer, her waggle runs crisp and repeated, as if she cannot help herself. The eaves scout dances too, but for fewer circuits per bout. The birdhouse scout dances briefly, then stops; perhaps she does not even return to dance again.

Followers respond. A few recruits go to the oak, a few to the eaves, perhaps one to the birdhouse. The recruits who visit the birdhouse quickly learn why it is a poor choice and return unenthused; they do not add much dancing. The oak recruits return energized and join the campaign. Now, because more dances point to the oak, more recruits are sent to the oak, and so on. The debate shifts without anyone being "outvoted" in a single moment. It shifts because attention is self-reinforcing when it is tethered to a real advantage.

This is the swarm's positive feedback: good options generate more advocacy, and more advocacy generates more inspection. Positive feedback is powerful but dangerous. In engines, it can cause runaway; in crowds, it can cause panic. The swarm tempers it with two quiet forms of negative feedback. One is the natural decay of dancing. The other is the finite supply of scouts. Every bee recruited to inspect one site is, for a time, not inspecting another. As a campaign grows, it consumes the swarm's investigative bandwidth, leaving less room for new proposals. This makes the system efficient when it is close to a decision, but it also increases the value of early diversity, when the swarm most needs broad sampling.

Some of the most revealing moments in honeybee house-hunting come not when the swarm is choosing between a good and a bad option, but when it is choosing between two good ones. Two excellent cavities can each inspire long dances and vigorous recruitment. The dance floor becomes crowded with competing maps, and the swarm teeters near a deadlock. It is in these cases that the details of the "debate" matter. Dance decay helps, because it allows one campaign to fade if it fails to keep renewing itself with returning supporters. In later phases, bees also employ more direct forms of competition, including signals that can interrupt other dances. But even before those more confrontational tactics become prominent, the basic rules of the dance floor already push the system toward resolution: enthusiasm must be maintained, recruits must return convinced, and time is always running.

If this sounds uncannily like human information ecosystems, that is because the underlying problem is shared: how to turn many partial, local observations into a single collective action without a leader who knows everything. Human groups often fail because we broadcast confidence rather than evidence, or because copying becomes easier than checking. Bees have copying—recruitment is a form of copying—but copying is tightly linked to verification. A follower does not become a full advocate by merely watching; she must usually fly out and inspect the site herself. On the dance floor, the swarm is not passing around a rumor; it is passing around invitations to go see for yourself.

That point can be sharpened with an old anecdote from the early days of bee science, when naturalists tried to understand the waggle dance as if it were a symbolic language meant for human eyes. The temptation was to treat the dance as a precise code and the followers as decoders. But the swarm's decision-making looks less like decoding a message and more like allocating labor. The dance points attention outward. It does not settle the matter. Each recruit is a mobile sensor added to the site. The swarm is, in effect, building a temporary measurement device out of flying bodies.

A striking feature of this device is how it handles disagreement without needing to eliminate it immediately. On the swarm, multiple dances can coexist. No authority silences the minority proposal at the outset. In fact, allowing several proposals to coexist is part of the method. The swarm is not embarrassed by disagreement; it uses it. Competing dances keep the swarm sampling multiple options until one option earns dominance by sustaining recruitment.

This "market of information" has another subtle benefit: it protects the swarm from overreacting to a single persuasive individual. A particularly energetic scout can recruit followers, but she cannot manufacture returning advocates unless the site itself supports that enthusiasm. If she dances wildly for a mediocre cavity, her recruits will tend to come back disappointed and stop dancing. Her campaign collapses under its own lack of reinforcement. If she dances wildly for a superb cavity, her recruits come back impressed and multiply her signal. Either way, the environment is allowed to answer back.

The relationship between site quality and dance strength is often described as a simple correlation, but it is more like an economic link between value and spending. Dancing is expensive. Time spent dancing is time not spent feeding, resting, or scouting elsewhere. A scout "pays" for the swarm's attention with her own effort, and she is willing to pay more for a site she rates highly. That is why the dance floor is not merely a place where information is displayed, but a place where information is *priced.* Better sites command higher expenditure.

The swarm's debate is therefore not a debate of speeches. It is a debate of commitments. Each dance is a public commitment to a particular location, expressed in repeated, strenuous movement. And because the dancers are embedded in a crowd of potential recruits, each commitment has the capacity to grow.

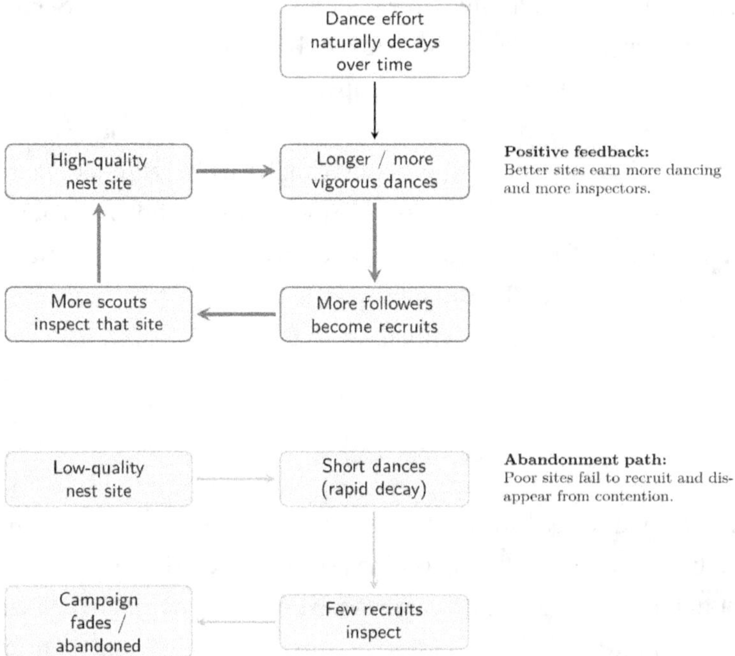

The elegance of the dance floor debate is that it solves two problems at once: it discovers options and it compares them. Discovery happens because any scout who finds something can put it into the swarm's shared arena by dancing. Comparison happens because dances are graded by the dancer's own evaluation and are allowed to compete for recruits.

A final detail completes the picture. The dance is not only a signal to others; it is a signal that the dancer herself can update. A scout may return to a site, reassess, and adjust her dancing accordingly. She may dance strongly, recruit others, then later weaken or stop if subsequent visits reveal flaws: dampness, an unexpected draft, an entrance too exposed. The swarm's "opinions" are therefore not fixed ballots but revisable commitments, continually pressured by reality.

Seen from the outside, the swarm's cluster is motionless. Seen up close, it is a roiling argument conducted in quiet, repetitive steps. A few bees turn their bodies into maps. Other bees press in to read those maps. Still others fly out to check whether the map is worth trusting. Without speeches, without a chairperson, without any bee understanding the whole landscape, the swarm builds an outcome that is hard to improve upon: a home chosen not by proclamation, but by a competition between proposals that must pay their way in attention and must survive contact with the world.

3.2 The Quorum Switch

A swarm can afford to be choosy only for so long. Hanging in a tight ball, the bees are living off the honey they carried out of the old nest in their stomachs. They are also exposed: a cold night can chill them into immobility, a wet day can sap their energy, and a curious bird can pick at the outer layer. The debate on the swarm's surface is therefore not a leisurely parliament. It is more like an emergency meeting held in the street, with everyone keeping one eye on the sky.

That urgency creates a puzzle. If the swarm's choice emerges from a competition among dances, why doesn't the process drag on indefinitely whenever there are several decent options? If the swarm simply waited until every dancer had stopped advertising all but one site, it could wait too long. And if it departed as soon as one site merely *looked* popular on the dance floor, it could leave on the basis of fashion rather than evidence. The swarm needs a stopping rule: a crisp, reliable signal that says, in effect, "Enough checking has been done; move now."

The stopping rule honeybees use is called *quorum sensing*. The phrase comes from politics and committees, but in biology it means something very specific: individuals change their behavior when they encounter enough of their peers in one place. Bacteria do it with chemicals, ants do it with body counts, and honeybee scouts do it with something that looks almost like a headcount conducted at the front door of a prospective home.

The key idea is simple. The dance floor is where proposals compete, but the decision is not finalized there. Instead, the swarm treats the candidate nest site itself as the ballot box. When enough scouts are simultaneously present at a particular site—not just one excited discoverer, not just a trickle of visitors over hours, but a dense little crowd—a switch flips in the scouts' behavior. They stop acting mainly as advertisers and start acting as mobilizers.

Before that switch, a scout returns from a site and dances, trying to recruit more inspectors. After the switch, scouts at the winning site begin producing a distinctive signal associated with imminent departure, and those signals spread through the swarm until the entire cluster becomes flight-ready. The swarm does not wait for unanimity on the dance floor. It waits for a threshold of physical commitment at the site.

Thomas Seeley and Kirk Visscher tested this idea in an unusually direct way in the early 2000s. If the swarm relied on a quorum at the chosen site, then anything that prevented any one site from accumulating enough scouts should delay the moment when the swarm "decides" and takes off. So they created conditions that encouraged scouts to spread their attention out among several nest boxes rather than converging quickly on one. The result was exactly what the quorum hypothesis predicts: when scouts were distributed across multiple boxes, the onset of the swarm's departure preparations was delayed. When a single box could more easily accumulate a crowd, the preparations began sooner. The swarm's stopping rule, in other words, did not seem to be "most dances point to one site," but "enough bees are *there*."

This is a subtle difference that matters. A dance is an advertisement, and advertisements can be plentiful even when the product has not been properly checked. A quorum is harder to fake. It requires that several bees, recruited independently and arriving at slightly different times, converge on the same location. It is evidence that the site has survived repeated inspection and repeated re-endorsement.

How many bees constitute "enough"? In Seeley and Visscher's observations, the threshold is often described as being on the order of a dozen or so scouts at the site at once—a figure commonly summarized as roughly 10 to 15 bees. That number is not a mystical constant. It can vary with circumstances: how difficult the site is to access, how restless the swarm is, how many scouts are active, how late in the day it has become. But it is striking that the swarm does not require hundreds. It requires a modest, achievable crowd: large enough to be meaningful, small enough to be reached before the swarm's fuel runs out.

The scouts do not stand in a line and count to twelve. Quorum sensing is implemented through local cues. A scout arrives at a candidate cavity and begins to explore. She bumps into other scouts. She hears them. She smells their footprints and the scent they leave as they investigate. She finds the entrance already being inspected. Density is hard to miss when you are a small animal in a small space. When that density crosses a threshold, the behavioral program changes.

One of the most memorable signs of that change is a sound: *worker piping*. A piping bee presses her thorax against other bees and produces a high-pitched buzz that lasts about a second. It is not the same as the familiar "queen piping" used in other contexts inside a hive. This piping is produced by workers, and many of them are scouts. Seeley and Jürgen Tautz described how piping begins at low levels roughly an hour before liftoff and rises toward a crescendo as departure approaches. The signal does not carry a map. It carries arousal. It tells the swarm to warm up.

Warming up is not a metaphor. Bees cannot fly cold. Their flight muscles must reach a working temperature, and in a cluster hanging in shade or wind, many bees will be below that. The swarm's body has to change state: from a hanging, energy-saving ball to an organism ready to launch into the air as a coherent cloud. Piping, along with jostling and brief bursts of activity, helps raise that readiness. In human terms, it is the difference between a crowd milling around a train platform and the moment when the doors open and everyone surges toward one direction. Without a coordinating signal, movement would be ragged and fragmented.

The remarkable part is how the swarm ties that arousal signal to evidence. It does not simply pipe because a dancer on the swarm surface is persuasive. It pipes when enough scouts have accumulated at one site. In experiments reported by Seeley and Visscher (2003), consensus among dancers on the swarm surface was neither necessary nor sufficient to trigger piping. The dance floor can remain somewhat pluralistic even as the swarm prepares to depart. That fact is easy to miss if one imagines the swarm making a decision the way humans do in a meeting: by first settling the argument, then acting. Honeybees overlap these phases. They can act while some dissent remains, because the stopping rule is not "everyone agrees," but "enough independent checkers are on-site."

A concrete scene makes this less abstract. Picture two nest boxes set out at the edge of a forest clearing. Both are acceptable, and both have their advocates. On the swarm, you could watch dances for Box A and Box B continuing side by side, perhaps with Box A slightly ahead. Now imagine standing near Box A. At first, visits are sporadic. A scout arrives, explores, leaves. Later, visits become more frequent. Two scouts overlap in time, then three. They cross paths at the entrance, one slipping past another. Inside, the air begins to carry the faint, accumulating odor of bees. A scout arriving now is not alone in her inspection; she is part of a growing crowd.

At some point, the site ceases to be merely a candidate and becomes a gathering point. Scouts linger. They re-enter and exit. The entrance becomes busy. This is what "quorum reached" looks like in the bee's world: not a number on a clipboard, but a qualitative shift in local bustle.

Now shift attention back to the swarm. The first pipers appear. They run through the cluster, pressing their thorax to other bees and producing those sharp one-second signals. More pipers join. The cluster loosens slightly; bees begin to fan their wings, to shuffle, to flex. If you could measure temperature across the swarm's surface, you would see it rising. The swarm, as a whole, is being pushed toward a new attractor state: flight.

One might ask why the swarm does not use a simpler cue, like "the strongest dance has been going on for a while," or "most dancers now favor one site." The trouble is that such cues are vulnerable to misleading social momentum. Early enthusiasm can be contagious. A charismatic dance can recruit followers who have not checked the site. Even the simple fact of being first can create a temporary lead. Quorum sensing shifts the stopping rule away from social display and toward physical verification. To reach a quorum, multiple bees must actually fly to the site, enter it, and remain there long enough that other arrivals encounter them. Quorum sensing thus compels the system to buy evidence with effort.

This also helps explain why the quorum threshold is modest. The goal is not to create an overwhelming majority at the site; it is to ensure a minimum level of verification and coordination. The swarm needs enough scouts present to do two jobs. One job is epistemic: confirm that the site is real, accessible, and suitable. The other is logistical: once the swarm takes off, it must be guided to the correct location. Scouts act as guides, streaking through the airborne swarm in the direction of the chosen home. A site with only one or two informed scouts is risky: the swarm could fragment, or lose the guide signal, or drift. A site with a dozen scouts is a different story. It has redundancy.

There is a human analogy here that does not trivialize the bees. Consider how a group of friends chooses a restaurant in an unfamiliar neighborhood. One person's confident recommendation may be wrong. Ten people independently checking reviews, menus, and walking past the door produce a more reliable decision. But the group cannot spend all night checking. At some point, enough evidence accumulates and someone says, "It looks good; let's go." In humans that sentence is spoken aloud and accepted socially. In bees it is embodied: enough scouts gather at one door, and the swarm's physiology begins to shift.

What makes this especially compelling is that the quorum is not a global measure. No bee knows how many scouts are at Box A and how many are at Box B. No bee needs to. Each scout is responding to local density where she stands. Local interactions, stacked together across many bees, generate a global decision. This is the same basic trick that allows other swarms to form lanes, patterns, or coordinated attacks without central direction. Here the emergent pattern is not a shape but a commitment.

It is tempting, when hearing about quorums, to imagine a single decisive moment: one scout notices the twelfth bee and declares victory. Real biological systems are rarely that clean. Thresholds are often fuzzy, and behavior changes can be probabilistic. One scout might begin piping at a density that another scout would still treat as "not yet." That variability is not a flaw. It can be a feature, smoothing the transition and preventing the system from oscillating. If every scout required exactly the same count, a brief fluctuation could flip behavior on and off. Variation makes the switch more robust.

The decision also remains, for a time, reversible in practice. If scouts begin to gather at one site but then discover a problem—a hidden crack that lets in water, a new occupant, a sudden draft as wind shifts—the buildup can stall. Without continued recruitment and presence, local density drops, and the mobilization signals weaken. The swarm's physiology does not sprint instantly from "debate" to "flight"; it ramps up. That ramp buys the system a little insurance against premature action.

A historical anecdote captures how surprising this all was when first studied carefully. For a long time, it was natural to assume that swarms must achieve near-unanimous agreement among dancers before departing, because that is how human committees try to avoid splitting. But careful observation undermined that intuition. Seeley and Visscher (2003) found that even when dances for multiple sites were still present, piping could begin, and the swarm could prepare to leave. Conversely, moments of apparent surface consensus could occur without triggering departure preparations. The swarm was not waiting for agreement on the dance floor. It was waiting for something else, something happening out of sight: a crowd at a particular door.

Once piping spreads, the swarm's body begins to behave as if the debate has ended, even if a few dancers keep advertising alternatives. Those late dances become irrelevant not because they are silenced, but because the swarm's attention has moved from deliberation to coordination. The system's priority shifts. A swarm that has warmed its flight muscles and begun to loosen its cluster is no longer in a stable state for continued argument. The organism is committing.

When liftoff finally happens, it can look abrupt: the cluster breaks, the bees pour into the air, and within seconds the branch is bare. But seen as the endpoint of quorum sensing, the takeoff is less like a sudden whim and more like the release of a spring. The energy has been building. Scouts at the chosen site have gathered and switched roles. Piping has spread through the swarm, raising readiness. The swarm has become, in effect, a single decision carried in thousands of muscles.

If the waggle dance is the swarm's way of arguing about possibilities, the quorum switch is its way of respecting time. It converts a graded, ongoing competition into a discrete, coordinated act. The swarm cannot keep thinking forever, and it cannot afford to stop thinking too soon. By anchoring the stop signal to the physical crowding of independent inspectors at a candidate home, the swarm's "enough" is neither purely social nor purely individual. It is a small, local fact that many bees can verify simply by being there, and that local fact is enough to make the whole living cloud finally let go of the branch.

3.3 Accuracy through Independence

A swarm's house-hunt has the drama of a close election and the stakes of a life raft. Yet the most impressive feature is not that a decision happens at all. It is that the decision is usually *good*. When observers first see the dance floor alive with competing proposals, it is natural to expect chaos: a few overconfident scouts, a rush of followers, a fashionable but flawed cavity winning on sheer momentum. Human crowds can behave like that, and human institutions spend enormous effort trying to prevent it.

Honeybees prevent it with a rule that sounds simple enough to be childish, but turns out to be mathematically potent: *don't repeat what you heard; check it yourself.* In bee terms, recruits do not become full advocates merely by following a dance. They typically fly to the site, inspect it, and only then decide how strongly to advertise. Recruitment spreads attention, but it does not automatically spread conviction.

This distinction between attention and conviction is the swarm's main defense against groupthink. Groupthink, in its simplest form, happens when individual judgments become correlated—not because everyone independently sees the same evidence, but because everyone is copying the same signal. Correlated errors are dangerous. If a single mistaken impression cascades through a population, the group's apparent confidence grows precisely as its accuracy collapses.

Bees use social influence, but they fence it in. Social influence answers the question, "Where should I look?" not "What should I think?" A dance can send you to a place you would never have found on your own. Once you arrive, the place itself must persuade you. If it does not, your advertising fades. If it does, you add your voice. In 2009, Christian List and Christoph Elsholtz modeled this combination of *interdependence* in sampling and *independence* in evaluation, showing that it offers a sweet spot: the swarm can pool information efficiently without allowing the same error to be amplified across everyone.

To see why that matters, it helps to imagine a world without this independence rule. Suppose recruits simply joined the dance they watched, like fans chanting a slogan they heard in a stadium. In that alternate swarm, the first energetic scout to return would create a self-reinforcing loop untethered from reality. Her dance recruits more dancers, who recruit more dancers, and soon the swarm is committed. The site might be excellent, or it might be a leaky cavity with a wide entrance and a resident colony of ants. The swarm's decision would depend heavily on who happened to arrive first and how vigorously she danced, not on the site's true value.

Real swarms do not behave like that. The recruit who visits a site does not merely confirm that the site exists; she evaluates it. She experiences its volume, its entrance, its smell, its dampness, its exposure. Then she returns and produces a dance whose intensity reflects her own assessment, not the enthusiasm she witnessed on the swarm. The swarm's information therefore comes from multiple, partly independent measurements of the environment.

"Independence" here is not absolute. Bees are not isolated rational agents. They can be biased by the mere fact that a site has attracted attention; a frequently advertised site gets more inspections, and more inspections increase the chance that some bees will judge it favorably. But the key is that each positive contribution must pass through a personal filter: a trip to the site. That filter breaks the pure copying chain.

Thomas Seeley and his collaborators found something else that makes independence even more effective: individual judgments are noisy. One scout may overestimate a site; another may underestimate it. Their dances will not be perfectly calibrated. Yet at the population level, the total dance activity directed at each site tends to track quality. Noise cancels when it is not correlated. This is the core intuition behind a famous idea in political mathematics, Condorcet's jury theorem, which says that if each voter is more likely than not to be right and their errors are independent, then a majority vote becomes increasingly reliable as the group grows.

Bees do not take a formal majority vote, but they exploit the same logic. Replace "voter" with "scout," and replace "vote" with "dance effort." A better site tends to produce stronger advocacy in more bees, and because those assessments are made after individual inspections, the errors are not perfectly aligned. The swarm's collective procedure, built out of recruitment and re-evaluation, acts like a distributed error-correcting code.

The power of that error correction becomes vivid when you consider just how limited a scout's information can be. A bee does not carry a blueprint of the ideal nest. She has evolved preferences and heuristics: prefer a cavity of a certain size, prefer a small entrance, prefer dryness and shelter. But she must infer these properties with crude sensors and limited time. Two scouts in the same box can experience slightly different drafts, notice different patches of moisture, or spend more time at one wall than another. Even the same scout can change her mind upon revisiting. Independence is the swarm's way of turning these imperfect impressions into a reliable collective signal.

The system also contains another subtle check on overconfidence: advocacy expires. A bee's dance does not continue indefinitely at full strength. It decays, sometimes over minutes to hours, unless reinforced by renewed visits. Seeley (2003) described how advocacy for losing sites tends to fade without requiring dramatic conversions, which helps prevent the swarm from remaining permanently divided. Dance decay is not only a mechanism for reaching consensus; it is a mechanism for forgetting stale or weakly supported information. In human terms, it is as if every claim came with an expiration date unless it was re-verified.

Independence in evaluation, decay in advocacy, and quorum sensing together produce a kind of statistical discipline. But there is still a lingering worry: what if two sites are genuinely similar? If both are good, each may generate strong, sincere advocacy after inspection. The swarm could deadlock: two large, stable camps, neither dwindling. A real swarm cannot afford that. The bees on the branch do not have the luxury of perfect deliberation.

Here honeybees reveal an additional, almost political sophistication: they are capable not only of praising a preferred option but of actively dampening support for alternatives. Seeley and Visscher (2012) showed that scouts can deliver "stop signals" to waggle dancers for competing sites. A stop signal is a brief vibrational pulse delivered by head-butting or bumping another bee, often accompanied by a short buzz. The effect is inhibitory: it causes the recipient to stop dancing, at least temporarily. Importantly, these signals can act across campaigns. Bees supporting one site can inhibit dances for another, functioning like cross-inhibition in neural decision circuits where one option suppresses the activity of another.

This may sound aggressive, even impolite. But it has a clear function. When two alternatives are closely matched, pure positive feedback can leave the swarm balanced on a knife-edge, with recruitment maintaining two stable peaks of support. Cross-inhibition helps break the deadlock and push the system toward a single outcome. It is not a replacement for independence; it rides on top of it. Stop signals are most useful when competing options are similarly attractive and the decision needs to become decisive.

A concrete example brings together these pieces. Imagine a swarm offered two excellent nest boxes, Box A and Box B, placed at similar distances. Both have good volume and a small entrance. The dance floor becomes crowded with intense dances for both. Recruits visit both and return with genuinely favorable assessments. Dance decay alone may not quickly eliminate one option, because both options keep earning renewed enthusiasm.

In that stalemate, the swarm is vulnerable to splitting. If half the swarm flies to Box A and half to Box B, both halves may fail. Honeybees avoid this not by insisting that every dancer agree before action begins, but by allowing the competitive dynamics to become sharper. As one option gains even a slight edge—perhaps by chance, perhaps by a subtle quality difference that only becomes clear after many inspections—its advocates can increase inhibitory pressure on the rival dances. The rival's recruitment slows. Fewer new inspectors arrive there, which reduces the chance of fresh enthusiastic dancers returning. The balance tips. The winner wins *decisively*, which is exactly what a swarm needs to avoid fragmentation.

What is striking is how these mechanisms resemble engineered decision systems, yet they arise from local rules. Engineers building a distributed sensor network face the same problems. If each sensor blindly forwards what it hears, the network can amplify noise and false alarms. If sensors independently verify before forwarding, accuracy improves but communication slows. If the network never stops gathering data, it never acts. If it acts too quickly, it acts on insufficient evidence. The honeybee swarm is solving this engineering problem using insects and evolution, not silicon and design meetings.

The bee's solution is also emotionally resonant, because it offers a hopeful counterpoint to the darker side of collective behavior. In human societies, independence is often treated as a moral virtue, but it is also a cognitive tool. When people stop checking and start copying, crowds can become irrational. When people check independently but remain socially connected enough to share where to look, groups can become surprisingly wise.

It is tempting to romanticize the bees as models of civic virtue, but the real lesson is more concrete and more interesting. The swarm's intelligence is procedural. It does not depend on any bee being especially smart, honest, or public-spirited. It depends on a set of constraints that make good outcomes statistically likely. If a bee wants to have influence, she must spend time dancing. If she wants her influence to persist, recruits must return convinced. If the swarm is to act, a site must accumulate a crowd of inspectors. If the swarm is stuck between two strong options, inhibition helps prevent permanent stalemate. None of this requires insight into the swarm's overall state; it requires only that each bee follow rules that tie her behavior to her own observations.

A historical aside illustrates how hard it can be to accept the power of such simple rules. For much of the twentieth century, the waggle dance itself was controversial, with arguments not only about what it encoded but about whether it could plausibly guide other bees. Once the dance's role in foraging became widely accepted, it was still natural to treat dancing as a form of instruction: one bee tells another where to go, and the second obeys. The house-hunting studies shifted that picture. A dance is less like a command and more like a pointer in a library catalog. It tells you where to find a book; it does not tell you whether the book is any good. That judgment is made by reading.

Independence also clarifies why swarms can be robust even when individual bees are wrong. A scout can misjudge volume, or fail to notice a crack, or be momentarily dazzled by a sheltered entrance that is actually too large. Her dance may still recruit attention, but the swarm does not hand her the keys on that basis alone. Her recruits become auditors. If the flaw is real, it will be discovered by at least some of them, and their reduced enthusiasm will blunt the campaign. The swarm is not immune to error, but it is designed to turn individual error into a small perturbation rather than a catastrophe.

There is, finally, a kind of humility embedded in the swarm's procedure. The bees do not behave as if any one evaluation is definitive. Dances decay. Scouts revisit. Recruitment continues until the evidence becomes thick enough at one site to trigger the quorum switch. Even then, the swarm's decision is not founded on the psychological satisfaction of agreement. It is founded on a physical accumulation of independent checkers at a particular place in the world.

If you stand near a swarm on the day it departs, you can sometimes feel the decision before you see it. The cluster becomes restless. A few bees begin piping. The surface loosens, and the hum changes pitch. Then the swarm lifts in a sudden, coordinated surge and streams away, guided by scouts that have already been to the chosen home often enough to know it in their muscles.

By the time the air fills with bees, the crucial work has already happened in a quieter form: repeated trips by small, stubborn individuals who refused to treat a dance as proof, and who insisted, in the only way a bee can insist, on going to look.

Chapter 4

The Wave That Watches

A massive flock of starlings turns instantly, seemingly by telepathy, yet the mechanism is far simpler and more elegant than a hive mind. By limiting their attention to immediate neighbors, individual creatures create a macroscopic system capable of transmitting information like a sound wave through a medium. We explore the physics of this living fluid, distinguishing between the myth of synchronized leadership and the reality of mathematical criticality.

4.1 The Algorithms of Flight

A flock of starlings can turn as if it were poured, not flown: a single, dark ribbon in the sky that flexes, folds, and reforms without tearing. From the ground it looks choreographed, even deliberate, as though some hidden conductor were tapping out the beat. Yet the most useful starting point is almost offensively simple: no one is in charge. Each bird has a body with limits, eyes with blind spots, and a brain that cannot possibly compute the trajectories of hundreds of neighbours. Whatever order we see has to be made out of small, local decisions, repeated many times per second.

That claim used to sound like a philosophical preference for decentralisation. In the late twentieth century it became an algorithm. In 1987 Craig Reynolds, working in computer graphics, wanted animated birds that moved like birds, not like a handful of dots dragged along a curve. His solution became famous under the name *Boids* (short for "bird-oid objects"), and it is still the cleanest demonstration that collective grace can come from individual thrift. Reynolds did not give his simulated birds a map of the flock, a leader to follow, or an agreed plan. He gave them three nudges, each based only on nearby neighbours. Out of that parsimony came a convincing murmur.

The three nudges are often named as if they were moral virtues: separation, alignment, cohesion. The names are helpful, but the deeper point is geometric. Each "boid" measures what is close enough to collide, what is close enough to copy, and what is far enough away to risk losing contact. The flock is not a thing above the birds; it is the pattern traced by these overlapping neighbourhoods, constantly recalculated. A turn is not a command issued from the centre. It is a chain reaction of small steering corrections, like a stadium wave made of wings rather than arms.

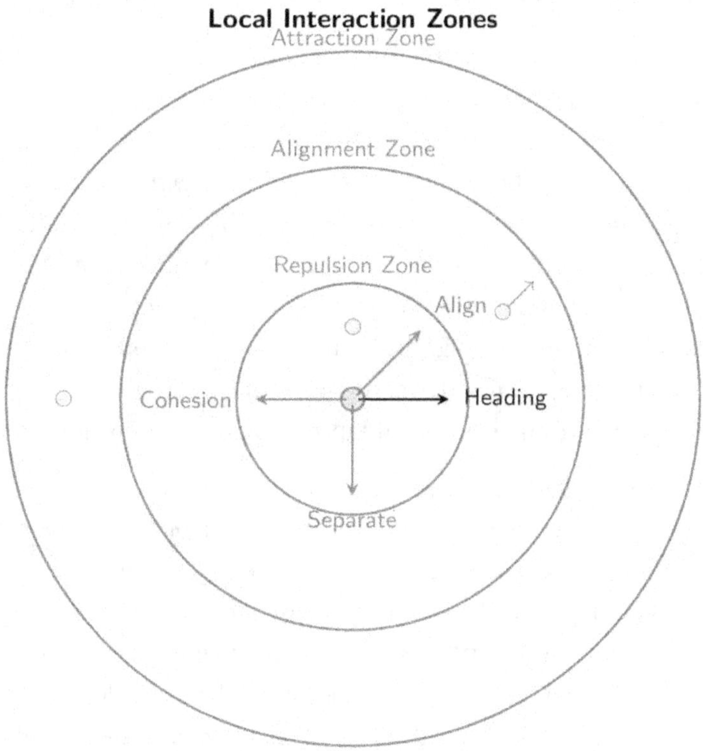

The picture helps because it makes a common misconception visible. People imagine flocking as a force of attraction, like gravity: birds pull toward the group, and somehow the group stays together. But gravity alone produces a stampede into collision. What stabilises the swarm is a balance between competing tendencies, with different "ranges" of relevance. Too close, and avoiding impact dominates everything else. At intermediate distances, matching direction matters most, because it reduces shear: you do not want to fly across your neighbour's path any more than you want to walk diagonally through a crowded corridor. Farther still, a mild pull keeps you from being left behind, a social tether that preserves membership without crushing the group into a point.

The genius of Reynolds' formulation is that it turns a biological puzzle into a small piece of control theory. Each rule can be expressed as a steering vector: a direction and strength added to the agent's current motion. If a neighbour is within the *zone of repulsion*, the steering vector points away from that neighbour. If neighbours fall in the *zone of orientation*, the agent adjusts its heading toward their average heading. If neighbours are only in the *zone of attraction*, the agent aims toward their average position. Summed together, plus some limits on acceleration and turning rate so the motion stays physically plausible, the result is not a rigid formation but a living compromise.

That word "compromise" matters because it prevents the rules from sounding like three separate behaviours stacked on top of one another. In practice they compete every moment. Imagine you are on a busy sidewalk. If a stranger suddenly steps into your path, you do not pause to compute a new "cohesion" goal with the crowd. You simply avoid collision. A half-second later, once the immediate danger is gone, you drift back into the flow, matching the general direction of movement. Birds do something similar in the air, only faster, and with the added complication that "up" and "down" are available. The recipe works because nature and algorithms share a preference: if a small calculation can keep you safe and coordinated, why perform a large one?

Separation is the least negotiable of the three rules because bodies cannot overlap. In a model it is often implemented as a strong, short-range repulsion: the closer the neighbour, the harder the push away. There is something visceral about it. The rule has the flavour of reflex, a shiver in the muscles, a quick widening of personal space. And it introduces the first hint that "order" can be an accident of constraints. If every agent avoids getting too close, then the group has a built-in incompressibility, a tendency not to collapse. The flock becomes, in a loose sense, a fluid with a maximum density.

Alignment is more subtle. Avoiding collision keeps you alive; aligning keeps you from constantly having to avoid collision. When headings are similar, relative speeds are small. Neighbours drift rather than cross. Local alignment reduces the turbulence a flock would otherwise generate in its own motion. One way to appreciate this is to imagine a group of cyclists in a city. If everyone points their front wheel in random directions, the group cannot move as a group at all; it becomes a pile-up. If most cyclists are roughly aligned, even without strict formation, the group can flow.

Cohesion is the rule that sounds most social, and in a way it is. It keeps the flock from evaporating. But in Reynolds' model it is not a mystical desire for togetherness; it is a response to isolation risk. A lone bird is easier prey. A lone simulated "boid" is also less stable, more exposed to noise. Cohesion is therefore a gentle corrective: if your neighbours are mostly to one side, you drift that way. Not too aggressively, because aggressive attraction would create collisions that separation then has to fix, producing jittery, nervous motion. The best cohesion is calm enough to be ignored whenever it conflicts with safety.

Those rules, stated plainly, can feel almost disappointing. Surely the sky-drama of a murmuration must require something more baroque than "don't crash, copy direction, don't fall behind." The disappointment fades when you watch what these rules do to information. A flock is not only a collection of bodies; it is a medium through which changes propagate. One bird's small deviation becomes its neighbour's small correction, which becomes another neighbour's correction, and so on. The flock begins to behave like a material that can carry a ripple, even though no bird "sends" a message in the way we usually mean it.

A concrete example makes the point. Suppose a peregrine falcon dives at the edge of a starling flock. The birds nearest the threat must both avoid collision with the falcon and avoid collision with each other as they swerve. Separation drives a sharp local expansion, creating space fast. Alignment then tries to keep neighbours moving in similar directions, so the expansion does not become a spray of individuals. Cohesion tugs the disturbed region back into the group once the immediate manoeuvre is over. The resulting shape change can look like a single coordinated turn, but it is built out of overlapping local necessities. In the language of physics, the flock has become an excitable medium: a small disturbance can trigger a moving front of change.

This is where Reynolds' origin in animation becomes more than a footnote. Computer graphics is unforgiving about motion that looks "too planned." Humans are extremely sensitive to the difference between a crowd and a parade. A parade has a script; a crowd has pressure, imitation, hesitation, and the occasional near-collision. Reynolds was trying to capture the latter, and in doing so he stumbled onto a general principle of swarm minds: complexity can be cheaper than you think. You do not need a blueprint for a cathedral if you can arrange for bricks to settle under gravity and constraints. You do not need a flight plan for a flock if you can arrange for each bird to react to what is nearby in a stable way.

Still, it would be a mistake to treat the three rules as if they were literally written in a bird's head. They are abstractions, like Newton's "forces" in a simple mechanics problem. Birds do not compute vectors; they move muscles. Their perception is not a perfect circle around them; it is shaped by the position of their eyes, the occlusion caused by other birds, and the fact that what matters most is often in front, not behind. This is why many later models replace neat concentric circles with sectors: a forward cone of attention, perhaps with a blind region to the rear. The spirit remains the same. The "algorithm" is whatever minimal set of local reactions can keep bodies separated, headings similar, and the group cohesive.

A second refinement comes from asking what, exactly, counts as a neighbour. In the simplest version of Boids, any agent within some metric distance contributes to your steering. That is an easy choice for a programmer, but it implies an odd biological burden. If the flock becomes denser, suddenly you have many more neighbours inside your radius; if it becomes sparse, you have fewer. Your brain would have to scale its attention with density, becoming a better computer precisely when the situation becomes more demanding. Empirical work on starlings, associated with the STARFLAG project, suggested something more plausible: birds seem to track a roughly fixed *number* of neighbours, about six or seven, rather than everyone within a fixed distance. The geometry shifts from metric to topological. The flock can compress or expand, but each bird's social "budget" stays constant.

That topological idea makes Reynolds' three rules feel less like a trick of simulation and more like an honest description of bounded attention. Think again of the sidewalk. You do not monitor everyone within ten metres. You monitor a handful of people who are most relevant to your immediate future: the ones directly ahead, the ones walking at your pace, the ones whose path might intersect yours. As the crowd thickens you do not suddenly track twenty faces; instead your attention becomes more selective. A fixed-neighbour rule also helps explain why flocks can remain coherent during dramatic shape changes. If cohesion depends on keeping links to specific neighbours rather than to everyone in a radius, then stretching does not necessarily snap the social fabric.

From here the three rules can be reinterpreted as three kinds of error-correction, each tuned to a different failure mode. Separation corrects the error of overlap. Alignment corrects the error of incoherent flow. Cohesion corrects the error of fragmentation. A good flocking regime is one in which the corrections do not fight each other too violently. If separation is too strong relative to cohesion, the flock shatters into spray. If cohesion is too strong relative to separation, the flock clumps and jitters. If alignment is weak, the group becomes a milling crowd, swirling without purpose. Somewhere between these extremes lies the region where a flock looks effortless: space is kept, direction is shared, and membership is maintained.

One of the most interesting consequences is that the "shape" of a flock is not an object; it is an argument. The edge birds experience different neighbourhoods than the interior birds. At the edge, cohesion has a clear direction: inward. In the interior, cohesion vectors largely cancel, leaving alignment and separation to do most of the work. This naturally produces something like surface tension. Edges are costly because they require sustained inward correction, and because individuals there have fewer neighbours on one side, making their local estimates noisier. Without any bird intending it, the flock tends toward shapes that reduce exposed edge relative to volume, especially when threatened. This does not require fear as an emotion in the model; it requires only that cohesion responds to missing neighbours.

The same logic scales down to much smaller creatures. A school of fish, viewed from above, often shows the same statistical dance: quick repulsion at close range, parallel headings at medium range, and a lingering attraction that keeps the school from thinning into solitary swimmers. The medium changes from air to water, but the algorithmic skeleton remains. Fluid dynamics adds additional coupling (fish feel vortices; birds ride wakes), yet the fundamental problem is shared: move without colliding, and do it in a way that keeps you hard to pick off.

What makes this worth calling a "swarm mind" rather than mere crowd dynamics is that the local rules turn perception into collective computation. Each agent does a small calculation from its limited sensory data. The group, by virtue of being many such calculators linked by motion, produces a coherent pattern that no single individual could predict or store. The pattern is not stored anywhere; it is enacted. It is a kind of memory without a memory palace, maintained only by continuous adjustment.

Seen this way, a murmuration becomes less like a single animal and more like a process: a rolling negotiation between three constraints. The flock survives by staying within a corridor of possibilities bounded by collision on one side and fragmentation on the other, with incoherence lurking as a subtler danger that makes both more likely. The remarkable thing is not that these constraints can be written as rules, but that rules so small can generate behaviour that looks, to a human eye, like intention. A wave of turning passes through the flock; each bird only needs to know what its neighbours are doing. The sky fills with order that belongs to no one and is owned, moment by moment, by all of them.

4.2 The Flash of Fear

A predator does not experience a flock the way an onlooker does. From the ground, a murmuration is a single expressive object: a veil pulled into new shapes by invisible hands. From the perspective of a peregrine falcon, it is closer to static on a screen. Hundreds of bodies, each reflecting light differently as wings tilt; edges dissolving and reappearing; gaps opening and closing before the mind can decide whether they are doors or traps. The flock's beauty is not incidental. It is part of its defence, because it rearranges the predator's main problem: choosing.

Predation is often described as a test of speed, strength, or stealth, but for many hunters it is also a test of attention. A falcon cannot strike "the flock" in general. It must select one bird, predict its future position, and commit to an interception. Fish-eating predators face the same constraint underwater, where a sudden swerve can turn a perfect line into a miss. When prey move collectively, they turn that constraint into a weapon. The term for this is the *confusion effect*: a dense, shifting group makes it harder for a predator to isolate, track, and successfully attack a single individual.

The confusion effect is not merely that "there are many targets." It is that the targets behave in ways that exploit the predator's own perceptual machinery. Visual systems, including ours, rely on continuity: tracking works by assuming that an object stays roughly itself from one moment to the next, and that it does not swap identities with its neighbour. In a tight flock or school, those assumptions become unreliable. Bodies overlap in projection; individuals exchange positions; a bird that was a half-second ago at the edge is suddenly inside; a fish that seemed isolated is swallowed back into the mass. The predator's best strategy can become to aim at the periphery, where tracking is easier. Many prey responses, in turn, can be understood as ways of denying the predator a stable periphery.

The three local rules that keep a group together in calm conditions are still there under attack, but they are driven hard enough to produce recognisable "moves," as if the swarm had a repertoire. Separation becomes explosive: create space now, even if it disrupts the neatness of the formation. Alignment becomes contagious: copy the neighbour's sudden turn, because hesitation is fatal. Cohesion becomes a repair mechanism: after the violent deformation, re-knit the group before individuals are left behind. Out of these amplified reflexes emerge patterns with names that sound almost theatrical: flash expansions, fountains, vacuoles, splits. They are not choreographies memorised in advance; they are the group's physics, revealed under stress.

Consider a fish school confronted by a fast, straight-line charge. A barracuda, for example, can accelerate quickly and strike like a thrown spear. If the school behaves like a timid crowd, compressing inward, it produces exactly what the predator wants: a thick block of bodies that increases the chance of contact. Many schools do the opposite. They *part*.

Observers have long noted a manoeuvre sometimes called the *fountain effect*. The predator drives into the oncoming school, and the fish divide around it, opening a narrow corridor along the attack path. For a moment the school resembles water around a rock, flowing around a hard object with surprisingly little chaos. Then, just as striking would require the predator to choose and commit, the corridor closes behind it. From above, it can look as though the school has performed a trick: the attacker passes through without "catching" anything, and the prey reform into a cohesive body, now behind the predator, with the hunter facing empty water.

This is where the confusion effect becomes mechanical. The predator's trajectory is usually constrained: it has momentum, a preferred turn radius, a plan to avoid injuring itself. The school's advantage is that it is made of many small bodies, each capable of a sharp turn, and each copying its neighbours fast enough that the group can change state on a timescale that competes with the predator's corrections. If the predator barrels forward, the safest action for an individual fish may be a sideways dart away from the strike line. But if only a few do this, they become isolated and easy to pick off. The fountain effect is a massed sideways dart that keeps isolation rare: separation pushes individuals away from the oncoming threat, alignment makes that push coherent (so they do not collide in their panic), and cohesion makes the re-closure swift.

The fountain effect is also a reminder that "defence" is not always the same as "dispersal." Many animals hide by scattering, because a dispersed group offers fewer clustered opportunities. But dispersal is dangerous if the predator can single out individuals in open space. The most protective response may be to deform without disintegrating: to become briefly porous, then regain density. A school that can open and close like a muscle is hard to exploit. It refuses the predator both the dense clump and the isolated straggler.

Bird flocks have their own versions of this logic, adapted to three-dimensional air and the risks of wing-to-wing collision. When a peregrine makes a pass at a starling flock, the group can perform a sudden radial "breathing out" that has been termed a *flash expansion*. The flock enlarges rapidly, thinning its density in a burst that lasts only a moment. To a predator relying on a last-second adjustment, this can be devastating: the planned interception point is suddenly empty. Immediately afterward, the flock may contract again, or fold into a different shape, as if it has re-zipped itself.

The word "flash" is apt in more than one sense. The expansion can generate a shimmer because the birds tilt in near synchrony. A starling's dark body and lighter underwing surfaces produce strong contrast as it banks. If many birds bank together, the group's appearance can change abruptly, like a screen flickering between images. In the predator's visual field, this is not just motion; it is a change in the very features used to track. A bird that was a dark dot becomes a light dash. Edges blur. The predator's attention, already strained by the density of targets, must now contend with a rapid alternation of brightness and orientation.

There is a long history of people noticing such collective "flashes" in fish as well. Early naturalists, and later fishermen, often described a school as a single reflective body that "turned silver" when startled. The language is anthropomorphic, but the phenomenon is real: many fish have reflective scales that act like mirrors, and coordinated turning can produce synchronized changes in reflected light. A predator trying to track a single fish in that glinting storm faces a problem similar to the falcon's: the signal is high-contrast and rapid, but not stable enough to lock onto.

A concrete historical anecdote comes from the Second World War, when engineers and military observers became intensely interested in deception by pattern. Radar and visual camouflage were obvious concerns, but so was the general question of how many moving objects can be tracked at once, and how quickly confusion sets in. The emerging science of human factors and attention had a grim impetus: pilots and gunners needed to pick out targets in crowded skies. While this research was not about starlings or herring, it established a broader point that animal groups exploit without knowing it: tracking many similar moving objects is hard even for a trained brain with good optics, and failure often arrives suddenly rather than gradually. The prey group's task is to push the predator's cognitive load over that cliff.

The most intriguing defence patterns look less like expansion or parting and more like *waves*. A wave is not an object; it is a travelling change. In a stadium, the "Mexican wave" is a moving band of raised arms, not a group of fans running along the seats. In a flock under attack, something similar can happen: a band of abrupt manoeuvres propagates through the group, often starting near the predator's approach and moving away from it. Observers sometimes describe a dark band sweeping across a starling murmuration, like a shadow sliding over cloth. This kind of travelling band has been called an *agitation wave*, because it marks a region where birds are banking more sharply or changing speed and direction more abruptly.

The visual impression is so strong that it invites mistaken explanations. One tempting story is that the band is a literal density change:

birds bunch up, creating a darker region, and that bunching travels like a compression wave through a gas. Another story is that the band is mostly optical: when birds bank, their wing surfaces present different areas to the observer, so a region of synchronized banking can appear darker or lighter even without major changes in density. Both mechanisms may operate, and researchers still debate which dominates in particular cases. What matters for the swarm mind is that, whichever mechanism produces the appearance, the *cause* of the propagation can be local copying: one bird's abrupt bank triggers its neighbours' bank, and so on. The wave can carry "fear" without any bird broadcasting a signal beyond its immediate vicinity.

Waves also have a crucial property: they can *dampen.* A ripple on a pond fades as it travels because energy is dissipated. Similarly, agitation waves in flocks can lose intensity with distance: the band becomes less pronounced, the manoeuvres less extreme. Damping is not a failure; it can be protection. If panic propagated undiminished through an entire group every time a predator approached, the flock would waste enormous energy and increase collision risk. A damped wave confines the most violent evasive behaviour to where it is needed, while still providing a quick, distributed warning. In everyday life, humans do something similar in crowds: a shout at the front can trigger a flinch behind, but only those closest to the disturbance leap aside; farther back, people merely pause, glance, adjust. The group remains coherent because not everyone overreacts.

This balance between responsiveness and stability is delicate. A flock that is too "stiff" fails to propagate the evasive turn quickly enough; individuals near the predator pay the cost. A flock that is too "loose" overreacts; it fragments, creating isolated victims. The confusion effect thrives in the middle ground, where the predator sees *change* everywhere but *separation* almost nowhere. The group becomes hard to parse into individuals, yet it does not surrender individuals into the open.

One way to make this more intuitive is to think about what a predator needs at the final moment of attack. It needs *predictability*. The predator's body is a ballistic instrument, especially in a high-speed dive or a sudden underwater lunge. Its strike is a commitment to a future position. Any prey behaviour that breaks predictability at the right moment can convert a near-certain capture into a miss. But the prey cannot simply move unpredictably as individuals; random motion would cause collisions, waste energy, and often make matters worse. The prey need a collective way of being unpredictable to the predator while remaining predictable to each other. That is exactly what local interaction rules provide. To a neighbour, your sudden turn is legible because it is constrained: you turn away from the threat, you avoid collision, you keep pace. To the predator, the resulting mass motion is illegible because it is high-dimensional: hundreds of constrained turns superpose into a changing geometry that cannot be reduced to "the target goes left."

Predators respond with their own counter-adaptations. Some specialise in edge attacks. Some use repeated passes to fatigue the group. Some attempt to force a split. The most dramatic failure of collective defence is fragmentation: a flock torn into pieces, a school divided into smaller schools, each now easier to exploit. Yet even splitting can be ambiguous. A split may sacrifice a few individuals but preserve the majority, much as a lizard's shed tail distracts a predator. In that grim arithmetic, the swarm mind is not benevolent. It is an emergent compromise between individual survival and collective dynamics, and those are not always aligned.

The emotional temptation is to romanticise the group as a protective family. The reality is colder and, in a way, more impressive. No fish in the fountain effect decides to "save" the others. It responds to a fast-approaching threat and to the immediate motion of its neighbours. No starling in an agitation wave chooses to "broadcast alarm." It banks because the birds beside it bank, because staying aligned reduces collision, and because cohesion makes it costly to peel off alone. The predator, in turn, faces not a council of prey but a physical process that has learned, over evolutionary time, to occupy the predator's attention, to distort its predictions, and to turn selection into hesitation.

Stand long enough near a winter roost and you can see this logic written in the sky. A peregrine appears as a hard point of intent, a line drawn through air. The flock answers with softness: a deformation that is everywhere local, nowhere commanded, and yet coherent enough to deny the predator the one thing it needs. For a moment the flock becomes a flash of fear made visible, and then, as the attack passes, the air fills again with the quieter work of staying together.

4.3 The Speed of Consensus

A flock's most astonishing feat is not that it holds a shape, but that it *changes* shape without tearing. The turn begins somewhere—a ripple in the edge, a few birds snapping into a new heading—and then, in less time than it takes to draw breath, the whole body of the flock has agreed. Agreement here does not mean a vote. It means something stricter and more physical: velocities have aligned. A thousand independent bodies have found the same direction quickly enough that none are left crossing paths. The result looks like intention, yet no bird has access to the whole picture.

That raises a question that sounds simple and becomes thorny when you try to answer it: how does information travel through the crowd? Not a spoken message, not a signal in the usual sense, but the practical information that *a turn is happening* and *which way to go*. In the previous stories the group's safety depended on producing confusion for a predator. Here the group's safety depends on producing clarity for itself.

If information spread only by individual imitation, one neighbour at a time, one might imagine it creeping through the flock like spilled ink: slow, smudged, weakening as it goes. That is diffusion, the same mathematics that describes how a drop of dye disperses in water. Many early models of collective motion predicted something like this. A few individuals change direction; their immediate neighbours adjust; the change spreads outward, but it attenuates, and distance dilutes it. Farther away, the message becomes faint, and the flock risks fragmenting into regions that disagree.

Real flocks do not behave like that. High-speed, multi-camera studies of starling murmurations in the early 2010s made the point vividly: when a flock turns, the change in direction propagates across it in a *wave-like* manner, not a diffusive blur. A wave has a speed. It travels without losing its identity as quickly. If you flick one end of a rope, a pulse runs down the rope; it does not slowly seep. The same is true of the flock's turning event: a moving front of directional change sweeps through the birds, maintaining its coherence as it goes.

To say "wave" is not to claim that birds are connected by some hidden elastic band. It is to claim something about the *rules of response*. In a wave, each part of the medium resists being changed too quickly, and that resistance (physicists call it inertia) couples neighbours so that a push becomes a travelling disturbance rather than a local wobble. Birds have inertia in the literal sense: they are masses in flight and cannot turn instantaneously. But the relevant inertia is also behavioural. A bird tends to keep doing what it is doing unless a sufficiently strong local reason tells it otherwise. That persistence can sound like a nuisance—surely you want a bird to be agile—yet it is essential for carrying a clear signal. Without some form of inertia, a local change would be immediately drowned in the flock's ordinary noise: small, continuous jitters in heading that come from turbulence, wingbeats, and imperfect sensing.

A concrete analogy helps. Imagine a crowded applause in a theatre. If each person claps at any tempo they please, the sound is a hiss. But under certain conditions, the audience can lock into rhythmic clapping: a pulse that sweeps through the room, a collective beat. That beat is a form of consensus propagated in time. It requires two ingredients. One is coupling: people hear their neighbours and adjust. The other is persistence: you keep a tempo for a moment rather than changing instantly with every stimulus. In the flock, coupling comes from local alignment; persistence comes from behavioural inertia. Together they allow a turn to travel as a recognisable event.

The speed of such consensus is not only a matter of reflexes; it is also a matter of *who counts as a neighbour*. The earlier picture of concentric zones is useful, but it hides an awkward detail: a flock expands and contracts constantly. Under predation it can stretch dramatically, producing regions of high density and low density side by side. If birds interacted with everyone inside a fixed metric radius, then in a dense patch each bird would have too many neighbours, while in a sparse patch it would have too few. The network of interactions would thicken and thin unpredictably. Messages could bottleneck. Cohesion could fail exactly when it is most needed.

Field studies of starling flocks, associated with the STARFLAG project and analysed by researchers including Ballerini and Cabibbo (2008), suggested a more robust arrangement: birds appear to interact with a roughly fixed *number* of neighbours, about six or seven, rather than everyone within a fixed distance. This is called a *topological* interaction, in contrast to a metric one. The word "topological" can sound forbidding, but it simply means that what matters is the pattern of connections—who is linked to whom—rather than the exact distances between them. If the flock stretches, your nearest seven neighbours may be farther away than before, but they are still your reference set. Your social attention is a constant budget.

That constant budget is plausible biologically. Tracking six or seven moving objects is hard but not absurd; tracking twenty or fifty is another story. Anyone who has tried to follow a child through a crowded playground knows the cognitive strain of object tracking. The topological rule has a practical advantage: it keeps the interaction network from dissolving when the group becomes uneven. A bird in a thin region does not become socially isolated merely because distances have increased. It still maintains links to a handful of neighbours, and those links are the highways along which turning information can travel.

The power of this idea is easier to feel if you imagine two extreme scenarios. In a metric world, a bird has a circle of influence: only birds inside a fixed radius matter. Stretch the flock, and that circle becomes empty. The bird is, for a moment, alone, and must decide without guidance. In a topological world, a bird has a roster: the closest few bodies, wherever they are. Stretch the flock, and the links lengthen rather than break. The flock can breathe without snapping.

This is not merely about staying together. It is about preserving *the speed of agreement* as the group deforms. A wave needs a continuous medium. A rope transmits a pulse because its fibres remain connected even when stretched. A topological interaction network is a kind of connection that persists under stretching. It does not guarantee perfect transmission, but it prevents the most catastrophic failure: local changes that cannot find a path.

Once you start thinking of the flock as a network, a new set of questions appears. How far does a local perturbation reach? Does the flock have a characteristic "communication distance," beyond which influence fades? Or can a small change, under the right conditions, affect the entire group?

Here the story takes a turn into ideas borrowed from statistical physics, the branch of physics that tries to describe crowds of many interacting parts. In magnets, for example, each atom has a tiny magnetic moment that prefers to align with its neighbours. At high temperatures the moments are disordered, pointing every which way; at low temperatures they can align, producing a magnet. The interesting behaviour happens near the boundary between these regimes, at a *critical point*. Near criticality, correlations can become long-range: a fluctuation in one region can be echoed far away. The system becomes exquisitely responsive, but also precarious.

Researchers including Cavagna and colleagues (2010) measured correlations in starling flocks and found something striking: fluctuations in flight direction appeared correlated across the entire flock in a scale-free way. "Scale-free" means there is no single characteristic length scale that limits the correlation. Instead, the correlation length grows with the size of the flock. In plain language, the flock behaves as though its "listening range" expands to match its own dimensions. If that is true, it helps explain why a turn can sweep through a large flock without fading into noise.

The temptation is to declare: the flock is tuned to criticality, like a magnet poised at its phase transition. The phrase has rhetorical power, and it captures a genuine scientific hypothesis. But it is not the only possibility, and the debate is revealing.

One reason criticality is attractive is that it offers a mechanism for sensitivity. Near a critical point, a system can respond strongly to small inputs. A handful of informed individuals could, in principle, tip the group. This resonates with theoretical work on collective decision-making by Couzin and Krause (2005), which showed that even a small minority with a preferred direction can guide a large group under certain conditions, without explicit signalling and without any individual needing to know who the "informed" ones are. The group can reach consensus because local coupling amplifies a weak bias into a collective outcome.

Another reason criticality is invoked is that some measured correlations in flocks seem difficult to explain as mere by-products of being ordered. In an ordered phase—a flock already flying coherently—physics provides reasons to expect long-range correlations in *direction* fluctuations because of symmetry: if all birds are roughly aligned, then small deviations can propagate with little cost. But researchers argued that correlations in *speed* fluctuations are harder to get "for free" from symmetry alone. Speed is not just a direction on a compass; it is a magnitude that costs energy to change. If speed correlations are also scale-free, that suggests a more special tuning.

Yet there are cautionary notes. Measuring correlations in a moving, changing animal group is not like measuring them in a quiet slab of material. The flock is finite, often strongly ordered, sometimes forced by external constraints (wind, predators, obstacles). A finite system can mimic critical-like behaviour without being precisely tuned to a critical point. Moreover, different models can generate long correlation lengths under a range of parameters. The upshot is not that criticality is wrong, but that it is a hypothesis with rivals, and the flock's elegance does not hinge on the outcome. The deeper lesson is that the group's interaction rules can place it in a regime where local perturbations are not trapped locally.

The wave-like spread of turning adds another layer. In 2014 Attanasi and colleagues analysed turning events and reported that the information about a change in direction propagates with a roughly constant speed and little attenuation, consistent with a wave. Standard alignment models, in their simplest form, often predict more diffusive behaviour. Matching the observed wave-like transport requires incorporating behavioural inertia: birds do not simply set their velocity equal to the average of neighbours; they have a tendency to resist abrupt changes, producing dynamics closer to an underdamped system than an overdamped one. In the applause analogy, it is the difference between someone instantly matching whatever rhythm they hear (which produces noise) and someone keeping a beat long enough for a pulse to travel.

The importance of inertia can be felt intuitively if you imagine a flock without it. Suppose each bird constantly updates its direction to the instantaneous average of its neighbours. In a turbulent environment with noisy measurements, that average jitters. Birds would twitch, and a turning signal would blur into ordinary noise. Inertia acts like a low-pass filter: it smooths the response, allowing coherent changes to stand out. But too much inertia would make the flock sluggish, unable to respond quickly to predators. The flock's dynamics therefore reflect a balance: responsive enough to manoeuvre, persistent enough to transmit.

At this point, the flock begins to resemble an engineered system. Engineers worry about communication networks that must stay connected under motion, about consensus algorithms that allow many agents to agree on a shared state, about signal propagation that must avoid both excessive damping (the message dies) and excessive amplification (the system oscillates or becomes unstable). The flock faces the same design constraints, but it solves them through evolution and physics rather than blueprints. Topological neighbour selection maintains connectivity under deformation. Local alignment provides coupling. Behavioural inertia preserves signal identity. The collective, taken together, becomes a material that can carry information as a wave.

A small but telling human example makes the abstract language of correlation and propagation less airy. Consider a crowd at a busy crosswalk when the light changes unexpectedly or a cyclist cuts through. People do not receive a verbal alert. Instead, a few bodies near the disturbance step back; their neighbours adjust; a ripple of posture and motion travels through the group. In a well-packed crowd, the ripple can move quickly and coherently, producing a collective pause. In a sparse crowd, it dies out. The difference is not intelligence; it is coupling and density. Replace feet with wings and the same physics can become a life-saving turn.

One might still ask: if the flock is so good at transmitting information, why does it not become rigid, like a single solid object? The answer is that the flock's "connections" are not bonds. They are fleeting, constantly renegotiated by local geometry. A bird's seven neighbours are not permanent companions; they change as the flock flows. This continual rewiring is a kind of lubricant. It prevents the group from locking into a brittle lattice, and it spreads information through multiple routes. In network terms, it increases redundancy: if one path is disrupted, others exist. In everyday terms, it means the flock can be both coherent and fluid.

There is also a more uncomfortable truth about consensus. Agreement is not always global. Under extreme stress, groups can split. Even in ordinary motion, local regions can briefly disagree, producing shear and folds. Consensus is a dynamic achievement, not a permanent state. The impressive fact is not that the flock never fragments, but that it so often avoids fragmentation despite the speed of the manoeuvres and the lack of central control.

Stand under a murmuration at dusk and it is hard not to feel that something like a mind is present. The scientific description does not drain that feeling; it sharpens it. The mind, here, is not a thinking thing hovering above the birds. It is the flock's capacity to move information through itself quickly and cleanly, turning local perceptions into collective action. The turn is the signature: a decision made without a decider, carried as a wave through a network of seven-neighbour attentions, held together by the quiet insistence of bodies that cannot change direction without consequence.

Chapter 5

Cartographers of Scent

Without eyes to see the landscape or commanders to direct traffic, swarms construct highly efficient logistical networks using nothing but volatile chemicals and local interactions. By exploring how individual choices cascade into collective geometries, we reveal the biological algorithms that solve the "Traveling Salesperson Problem" far more elegantly than brute-force computation.

5.1 The Geometry of Rumor

A lone ant leaves the nest with no map, no instructions, and no clear destination. She moves the way a fingertip searches for a light switch in a dark room: forward, pause, adjust, try again. To a human eye it can look like indecision, but to the colony it is reconnaissance. Most of these walks find nothing. A few end in sugar, a crushed insect, a droplet of honeydew. The surprise is not that an ant can stumble into fortune; it is that a thousand stumbles can be braided into a reliable road.

The braid begins with a chemical trace so faint it might as well be a whisper. Many ants lay pheromone as they walk, a deposit from glands that can be released in tiny pulses. Pheromone is not a command in the way a shouted order is a command. It is more like a rumor left in the environment: *I went this way and did not immedi-*

ately regret it. Another ant crosses the same patch of ground minutes later. She does not read the rumor and obey; she merely becomes slightly more likely to drift in that direction. That difference can be vanishingly small. Yet with enough walkers, and with the right geometry, those small differences do not stay small.

One way to see the power of this is to imagine the simplest problem a foraging colony faces: two routes to the same food. A person would treat the choice as a matter of comparing lengths, then committing to the shortest. Ants do not measure length with rulers or odometers. Their legs do not click off meters in a tidy internal register. If a colony ends up using a shorter path, it is not because individuals have solved a geometry exercise. It is because the colony has invented a feedback loop in which time and traffic do the measuring.

The classic demonstration is the "double bridge" experiment, associated with the Brussels school of collective behavior (Jean-Louis Deneubourg and collaborators) and later repeated and refined many times. The setup is deliberately modest. Place a nest on one side, food on the other, and between them two alternative bridges that reconnect the same endpoints: one short and one long. At the start, both branches are clean. The first explorers reach the junction and split by chance. Some take the short bridge, some the long. A few minutes later the symmetry is gone. A thick line of ants streams along one branch, while the other becomes a minor road, then a neglected footpath. Most often the winner is the shorter bridge.

The trick lies in a timing advantage so ordinary it is easy to miss. An ant returning from food to nest lays pheromone as she goes. The faster she completes a loop, the sooner she reinforces the path again. A short route is a fast loop. Fast loops mean more deposits per unit time. More deposits make the route more attractive to the next ants arriving at the junction, who then choose it slightly more often, making the traffic denser still. Density is not just the number of ants on the path; it is the number of opportunities per minute for pheromone to be refreshed. That is what turns distance into chemistry.

This is positive feedback in its most compact form: a small lead becomes a larger lead because the lead itself accelerates the process that created it. Economists describe a similar process in markets when popularity attracts investment that creates more popularity. Linguists see it when a slang term becomes fashionable, increasing its chance of being heard and repeated. The ant colony's version is less sentimental but more legible. The environment stores the rumor, and every passing ant becomes both listener and broadcaster.

There is a further twist, discovered when researchers paid close attention to what individual ants actually do on the bridge. It is tempting to assume that pheromone is laid in a continuous stripe and that ants then simply follow the darkest stripe. But foragers are not train cars locked to rails. They hesitate, overshoot, correct, sometimes turn around. In experiments with *Lasius niger*, Beckers and Deneubourg (1992) found that U-turns can amplify the advantage of the shorter branch. A longer branch gives more time and more space for ants to drift, hesitate, and reverse, which changes where pheromone gets concentrated and how often ants re-encounter the decision point. In other words, geometry does not merely change travel time; it changes behavior mid-route, feeding back into how strongly the route is reinforced. The path is not an inert corridor. It is a behavioral machine.

Once the idea is clear, the double bridge starts to resemble a simple computer program distributed across hundreds of tiny bodies. Each ant executes a rule of thumb: *walk, sample, bias toward pheromone, add a little pheromone as you go.* The environment acts as memory: recent history remains as scent for minutes or hours, fading by evaporation and being overwritten by new deposits. The colony's "calculation" is not a calculation in the sense of representing numbers. It is iterative trial, continuously corrected by fresh experience. A route that speeds up returns tends to be strengthened; a route that slows returns tends to be weakened. Distance is translated into reinforcement rate.

Rumor is an especially fitting metaphor because the process does not require accuracy in any individual report. The first ants do not need to know which bridge is shorter. They do not even need to have found food for the initial asymmetry to begin: a few slightly more frequent crossings in one direction can be enough to nudge subsequent choices. And because the response to pheromone is often nonlinear—a strong trail can trigger disproportionately stronger following than a weak one—the system is primed for "symmetry breaking." Two identical options need not remain equally used. Randomness is not an obstacle here; it is the spark that allows a decision to crystallize.

That nonlinearity matters. If ants responded in a perfectly proportional way—twice the pheromone, twice the likelihood—then the colony might merely mirror initial conditions forever, drifting without committing. A nonlinear response curve acts like a social tipping point. Below a certain strength, a trail is background noise; above it, it becomes a magnet. In the laboratory this can look like a sudden phase change: one moment a scattered stream, the next a highway.

A historical anecdote makes the same point from a human angle. In the early twentieth century, as cities grew and department stores multiplied, shopkeepers learned that "traffic" could be engineered without ever giving customers explicit routes. Place a popular display in a particular spot, and people cluster there. The cluster itself draws more people, who then linger and create still more cluster. The original attraction can be modest; the feedback is the point. Architects and retailers turned corridors, doorways, and sightlines into instruments for amplifying attention. There is no pheromone in a mall, but there is an equivalent: bodies in space acting as cues. Ants discovered the same principle without architects. They use chemistry where humans use eyes and imitation. Both systems exploit a simple fact: a crowded path is information.

The double bridge also shows why the colony can be both decisive and wrong. If the long bridge happens to gain an early advantage—perhaps the first ant that finds food returns by that route and lays a stronger initial trace—the colony may lock into the longer route, at least temporarily. This is not a failure of the mechanism; it is part of the bargain. Positive feedback is a fast decision-maker, not a cautious judge. Once a trail becomes dominant, it recruits more ants, which strengthens it further, which recruits more ants. In stable conditions, speed is valuable. In changing conditions, speed can become stubbornness unless there is some way to counterbalance the amplification. (That counterbalance may take the form of pheromone decay, crowding effects, or explicit inhibitory signals, depending on the species and context.) The colony's mapping system is therefore best understood as a tug-of-war between reinforcement and forgetting.

Geometry enters again when the fork itself is examined. At a bifurcation, an ant does not hover like a drone, sampling both air currents and computing a weighted average. She walks, and the shape of the junction biases her body. If the branch she is already aligned with continues smoothly, it is easy to stay on it. If switching requires a sharp turn, it is harder. Jackson and Holcombe (2004) showed that in Pharaoh's ants, the angles of trail bifurcations can help ants reorient correctly when they have joined a trail from the "wrong" direction. Around a certain angle (roughly sixty degrees in their experiments), ants were more likely to correct their direction. This is a remarkable form of information storage: the geometry of the network is not just a consequence of traffic; it can become a signal that helps traffic behave sensibly.

That insight complicates the usual picture of pheromone trails as if they were mere lines of concentration. A trail is also an arrangement of bends, junctions, and branch angles. Those shapes influence encounter rates, turning probabilities, and how quickly ants return to reinforce. The landscape becomes part of the algorithm. A narrow passage channels ants into frequent contact, which can increase alignment and reduce wandering. A wide open area allows trails to diffuse and split. Even a small obstacle can act like punctuation in a sentence, marking where decisions are likely to be made again and again.

The reinforcement cycle can be told as a story with a protagonist who never learns. Imagine an ant who finds food by luck and hurries back to the nest. She lays pheromone, not because she knows she is communicating, but because that is what her physiology does when she is fed and moving. She exits again and reaches the same fork as before. She does not necessarily remember the precise turns; experiments suggest individuals need not rely on personal route memory for the colony to converge. She samples the ground and is nudged toward the stronger-smelling branch, which may now be the same one she took previously, but for reasons that have nothing to do with autobiography. A second ant reaches the fork moments later and is nudged the same way. Soon a third. Within a short time the trail is not a whisper but a rumor with witnesses.

At this point, it helps to be concrete about the "measurement" that the colony is performing. Suppose the short branch takes 30 seconds to traverse and the long branch takes 60 seconds. If each returning ant deposits pheromone at a roughly similar rate per step, then each ant on the short branch completes two full reinforcement loops in the time an ant on the long branch completes one. Even if the number of ants initially choosing each branch is equal, the short branch is refreshed twice as often. Because pheromone evaporates, refresh rate matters as much as total laid. The short branch maintains a higher average concentration. The ants arriving at the fork are not seeing distance; they are seeing the integrated history of recent returns.

The diagram below captures the logic. The key is not that the short branch is inherently more attractive. It is that it becomes more attractive *because* it is short, through a simple loop of time, deposition, and biased choice.

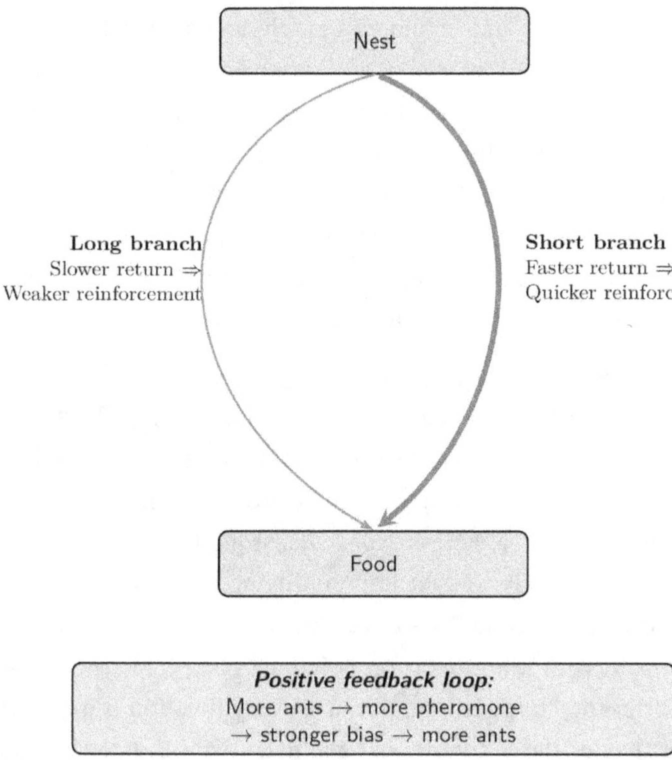

The elegance of this "collective computer" is that it uses the cheapest possible signals. Pheromone is not a detailed message with coordinates and annotations. It is scalar: more or less. Yet scalar signals can solve surprisingly rich problems when combined with the right feedback. The double bridge is a toy problem in the laboratory, but it hints at a general principle: if the world can store traces of behavior, and if agents can both read and write those traces locally, then a group can discover efficient routes without any member representing efficiency.

This principle has traveled far beyond entomology. In the 1990s and early 2000s, computer scientists borrowed ant trail logic to create "ant colony optimization," a family of algorithms for hard routing problems. A digital "ant" explores a graph, lays a virtual pheromone on edges it used, and later ants are biased to use edges with more pheromone. Over many iterations, good solutions tend to be reinforced. The striking thing is not that engineers were inspired by ants; inspiration is common. The striking thing is that the analogy holds because both systems exploit the same mathematics of reinforcement: exploration plus bias plus evaporation produces a search that is both adventurous and capable of settling.

But the biology remains more interesting than the metaphor, because real ants are optimizing under constraints that the tidy shortest-path story can obscure. A colony does not merely want the shortest line on a blank page. The world is uneven, dangerous, and noisy. A slightly longer route may be safer, less exposed, less likely to be washed away by rain, or less likely to split into confusing branches. In arboreal ants, where the "ground" is a three-dimensional graph of twigs and vines, a path with fewer decision points can be preferable to a path with shorter total edge length, because every junction is a chance to get lost. Under those conditions, geometry is no longer Euclidean distance; it is the structure of choices.

Even on flat ground, small physical details matter. A short bridge that forces ants into tight curves may increase collisions or slow movement enough to shrink its advantage. A long bridge that is straight and smooth can sometimes compete surprisingly well. That is why the double bridge is so useful: it strips away many of these complications to expose the core reinforcement logic, then invites the complications back in as questions rather than assumptions.

For a reader raised on the idea that intelligence must involve an internal model, the ant trail can feel like a cheat. Where is the representation? Where is the plan? The answer is that the representation is not in a head; it is in the world. The trail is a public memory written in a substance that fades on its own schedule. It is updated continuously by bodies moving at their own pace. The plan is not a blueprint; it is a tendency, a bias field that makes some futures more likely than others. When the colony "chooses" a path, the choice is not a vote taken at a meeting. It is a shift in probability at a fork, compounded thousands of times.

There is also a lesson here about rumor itself. In human societies, rumors can spread because they are exciting, not because they are true. They are amplified by repetition and by the visibility of repetition: the fact that many people have heard something becomes part of why it is believed. The ant colony's rumor is different in content but similar in mechanism. A trail becomes strong partly because it is already strong. That circularity is not a bug; it is the engine. The colony harnesses it for a practical end: moving food and information quickly through a landscape that offers no signage.

One can watch a foraging trail and feel that it is purposeful, almost stubborn, a line drawn with conviction. Underneath is a more delicate reality: the line persists only because it is continually rewritten. Stop the traffic and the "decision" evaporates into the air. Send the food elsewhere and, without countervailing forces, the colony may keep investing in an old rumor for a while, like a town that keeps driving to a market that has quietly closed. Yet this fragility is also what allows adaptation. A trail that is nothing but a chemical trace can be erased without grief. The colony is not attached to maps; it is attached to flow.

A final detail brings the story back to geometry in the most literal sense: the path chosen on a double bridge is not simply the shortest *in space*, but the shortest *in time*. Distance matters because it usually predicts time, but time is what the reinforcement loop actually measures. In that sense, the ants are cartographers of a peculiar kind. Their maps are not drawn in ink; they are drawn in rates. They discover, by iteration, the corridors where return is quickest and reinforcement is easiest. What looks like a line on the ground is really a loop in motion, a rumor that has found a shape sturdy enough to keep being told.

5.2 Paving with Kin

A column of army ants looks, from a distance, like a single animal poured onto the forest floor. It has a head—the leading edge that probes ahead—and it has flanks that ripple and fold around stones and roots. If you kneel close you see the illusion is built from thousands of bodies moving in a tight choreography, each ant a flickering unit in a larger flow. The strangest moment comes when the flow meets a break in the world: a gap between sticks, a small crevice in bark, a pothole in leaf litter, the sort of interruption you would step over without thinking. The column does not step over. It reorganizes itself, and sometimes the "ground" itself begins to move.

Army ants of the genus *Eciton* are famous for raids: sweeping, collective hunts that can flush insects, spiders, and even small vertebrates from hiding. The raid is a transportation problem as much as a predation problem. Prey must be carried back to the swarm or bivouac quickly, and the ants doing the carrying are also the ants needed to keep the raid going. Every second lost to detours or jams is paid for twice: in missed prey and in time exposed to predators and weather. The colony has no engineer with a clipboard, but it does have a remarkable ability to convert worker bodies into temporary infrastructure—living gravel, living guardrails, living bridges.

This sounds like a fairy tale until you see the mechanics. When traffic runs over uneven ground, ants at the bottom of a small depression can stop and brace themselves. Others run over them, using their bodies as a smoother surface. In a narrow crack, ants can link legs and mandibles, forming a span that carriers can traverse with fewer slips. The ants doing this do not look heroic. They simply stop in places where stopping is useful, and they remain stopped long enough for the benefit to accumulate.

The puzzle is not that an ant can cling to a neighbor. Many social insects can form chains. The puzzle is how the colony decides *where* to do it and *how many* bodies to commit. A bridge that is too small fails and wastes time. A bridge that is too large succeeds but ties up workers who could be foraging, carrying prey, or defending the column. The colony must find something like a bargain price for stability: enough structure to keep traffic fast, not so much that the raid starves itself of labor.

The easiest way to think about this bargain is to imagine the raid as a river. A river that runs fast can carry sediment and oxygen; a river that is forced through a boulder field loses energy in turbulence. Army ants face a similar problem, except their "water" is made of discrete, decision-making bodies. When the ground is smooth and the column is unconstrained, individuals can afford small navigational errors. A carrier can step around a pebble or pause for a fraction of a second without consequences. As the substrate becomes rougher, those small pauses begin to couple. One ant hesitates, the one behind bumps into her, the one behind that swerves, and suddenly a local slowdown becomes a knot.

The colony's response is an architectural one: reduce the roughness by filling it.

One of the most striking studies of this behavior came from observing *Eciton* columns on natural terrain and then reproducing key features in controlled setups. Researchers provided obstacles, gaps, or depressions of different sizes and watched how ants responded. Without needing to "measure" in any human sense, the ants tended to allocate more bodies to fill larger gaps and fewer bodies to fill smaller ones. The construction also adjusted over time: if traffic increased, bridges could thicken; if traffic decreased, bridges could shrink as ants peeled away and rejoined the flow. The structure behaved less like a monument and more like a variable resistor in an electrical circuit, changing its properties in response to current.

A living bridge is not only a span; it is a decision about who becomes ground and who remains traveler. Becoming ground has costs. The ant in the bridge is immobilized. She is also exposed. A bridge is in the open, often near the leading edge of a raid where predators may pick off individuals. The bridge ant may be stepped on and jostled for minutes. Yet the cost is distributed. No ant needs to volunteer repeatedly. The bridge is made of replaceable parts, and the colony's stability comes from not depending on any particular part for long.

The simplest local rule that could create this is also the most plausible: stop when you are experiencing repeated collisions or when the terrain forces your body into a stable bracing position; remain stopped if you continue to be useful, leave if not. That may sound circular, but it is exactly the kind of circularity that swarm systems exploit. "Useful" is not a concept an ant needs to represent. Usefulness can be measured indirectly through contact rates: if many ants pass over you, you stay; if few do, you go. The living bridge then becomes a self-tuning structure, built by local pressure.

A historical anecdote helps make the logic intuitive. Before modern road engineering, many rural roads were improved not by pouring asphalt but by "corduroy" construction: laying logs across muddy stretches so wagons would not sink. The logs were costly in labor and timber. Build too little corduroy and the wagons bog down. Build too much and you waste resources that could have gone into fields or fences. Communities learned to lay corduroy precisely where traffic demanded it, often seasonally. Army ants have converged on a similar strategy with a different material: instead of logs, they lay down kin.

Kin matters here in two ways. The first is obvious: the ants in the column are members of the same colony, and the benefits of any individual's sacrifice return to relatives. Natural selection can favor costly acts when they help genetic kin survive and reproduce. But kin also matters in a second, more immediate, chemical sense. The colony is held together by a shared odor profile dominated by cuticular hydrocarbons (CHCs), waxy compounds coating the exoskeleton. These chemicals help ants recognize nestmates, typically at close range through antennal contact, and in some species they can be detected even without direct contact. A raid column is a dense stream of encounters; antennae constantly tap bodies, legs, and the substrate. The column is a moving cloud of colony odor.

That odor does not merely police who belongs. It can influence how ants behave in traffic. In crowding experiments with *Lasius niger*, cues resembling nestmate CHCs presented on encountered objects have been shown to modulate pheromone-laying, linking "who is near me" to how strongly positive signals are broadcast. The same principle can be applied to structural behavior: contact with nestmates is not just noise, it is data. A high rate of nestmate encounters is a proxy for density and urgency. Where density is high, the benefit of smoothing the road rises; where density is low, immobilizing workers is less justified. In that sense, kin odor is part of the sensory channel that tells an ant whether she is inside a critical stream of colony activity.

The living bridge also clarifies something easy to miss when thinking about pheromone trails: a trail is not only a line of chemical preference. It is a physical system that must carry loads. Pheromone can encourage more traffic down a particular route, but more traffic increases wear, congestion, and the probability of bottlenecks. Army ants are forced to confront these trade-offs because they often move in dense, two-way streams—outbound hunters flowing away from the bivouac and inbound carriers returning with prey. A narrow pinch point can become a "tax" on the entire raid. If a bridge shortens the path across a gap, it can reduce travel time, but if the bridge is too narrow it can also become the pinch point itself.

This is where the architecture becomes subtle. The best bridge is not necessarily the shortest; it is the bridge that maximizes throughput. Throughput is the number of ants (or prey items) transported per unit time. A bridge that shortens distance but forces single-file crossing may reduce throughput compared to a slightly longer detour that allows multiple lanes. In other words, the colony's objective is not only minimizing distance but managing flow under constraints.

One way to picture the trade-off is to imagine a gap of width w that can be crossed by a living bridge. Building a bridge uses N ants, which are removed from the pool of mobile workers. The payoff is that the distance traveled per trip decreases, increasing the rate at which each remaining carrier can complete a loop. Let the raid's effective carrying capacity be something like

$$\textit{foraging efficiency} \approx (\text{number of mobile workers}) \times (\text{trips per worker per time}) \times (\text{success per trip}).$$

A bridge decreases the first term and increases the second. For small gaps, the increase in trips per time can dominate, so building is beneficial. For large gaps, the number of ants needed may become so high that the colony cannibalizes its own workforce. The existence of a "sweet spot" is almost inevitable, even if no ant knows it exists.

In practice, the sweet spot is discovered through local feedback. If the gap is small, a handful of ants can stabilize it. Traffic over those ants is heavy, reinforcing their tendency to remain. If the gap is large, initial bridge-building may be sparse or unstable; ants may detach because few can cross, leading to an unhelpful structure that dissolves. Alternatively, a large gap may recruit many ants into the bridge, but as the mobile stream thins, traffic declines, causing the bridge to shrink. The colony hovers near an equilibrium where the bridge is just large enough to be worth maintaining.

The equilibrium is not purely chemical. It is also mechanical. A bridge ant experiences forces: tugs from neighbors, weight from passing ants, vibrations from the stream. Mechanical stability becomes a cue. If you are anchored in a tense, loaded structure, you are likely part of something functional. If the tension drops, you may be part of an unnecessary appendage. There is an appealing analogy here to how bones remodel in vertebrates, thickening under load and thinning when load is absent. Army ants are doing something like remodeling too, except the "tissue" is a temporary assembly of individuals.

The first naturalists who wrote about army ants were struck by their apparent discipline. Thomas Belt, an English naturalist in Nicaragua in the 1870s, described columns of raiding ants and the commotion they caused among other animals; similar accounts appear in the writings of Henry Walter Bates and others exploring the tropics. Their prose can read as if the ants were a military unit with officers and ranks. The living bridge complicates that image in a useful way. A bridge is disciplined without being commanded. It is coordinated without being planned. The "order" that observers admired is an emergent property of many ants responding to local pressures and cues.

The human temptation is to ask which ants are assigned as bridge-builders. Are there specialists? In many colonies there are size differences and task biases, but the living bridge points toward a more fluid organization. Any sufficiently typical worker can become infrastructure if she arrives at the right place under the right conditions. The colony does not need to maintain a special caste of engineers. It needs only the capacity for workers to switch roles quickly and reversibly. That reversibility is essential. A bridge that could not be dismantled would be a trap, permanently locking workers into a place that might soon become irrelevant.

This flexibility depends on recognition and tolerance. If an ant is going to walk over another ant's body, grip her, or use her as a rung, she must treat her as part of the same system rather than as an intruder. Nestmate recognition via CHCs supports this tolerance. It is also a reminder that cooperation has a boundary. An army ant column is generous to its own and ruthless to outsiders. The same sensory machinery that enables smooth traffic can also trigger aggression when the odor is wrong. Diet can shift CHC profiles in some species, and laboratory manipulations have shown that changing what ants eat can change who they treat as kin, though field-oriented work suggests diet alone may not explain the full complexity of colony structure in invasive species like Argentine ants. In a raid, where chemical identity is constantly sampled through antennal contact, maintaining a coherent colony odor is part of keeping the infrastructure reliable.

It is hard not to feel a flicker of discomfort at the phrase "paving with kin." The image is of bodies used as tools, individuals flattened into a road. Yet it is also a reminder of scale. For the colony, the unit that persists through time is not the individual ant but the superorganism:

the reproductive queen(s) and the worker force that supports them. A worker's life is short. Her value is realized through the flow she helps sustain. From an evolutionary perspective, the bridge is not a tragedy; it is a strategy that has been tested against the harsh arithmetic of survival.

And the strategy is not blind self-destruction. Bridge ants do not simply die in place. They can leave. They can be replaced. The infrastructure is alive, responsive, and, in a literal sense, self-healing. If a predator snaps up a handful of ants from a bridge, the structure can sag and then thicken again as more ants stop and brace. If a fallen leaf shifts and closes the gap, the bridge dissolves and its ants rejoin the stream. The colony's architecture is a continuous negotiation with the environment.

Watch a raid long enough and you begin to see that the "road" is not always a line. It can widen into a fan when the prey field is rich, then narrow into a trunk when carriers return laden. It can split around obstacles, then rejoin. In those splits, subtle biases can be amplified just as pheromone biases are amplified at a fork. A slightly smoother branch draws more traffic; more traffic encourages more smoothing; the branch becomes dominant. The same positive feedback that can select a short bridge in a lab can select a passable corridor in a forest, with the additional twist that the corridor is being physically modified by the ants as they go.

There is a quiet, practical lesson in this for human design. We often treat infrastructure as something that must be planned in advance: roads, bridges, pipelines, communication networks. Army ants show another option: infrastructure that is *conditional*, built only where the current flow demands it, dismantled when the demand fades. It is difficult to do with steel and concrete. It is easier to do with modular, mobile components, which is why engineers building adaptive systems—robot swarms, self-reconfiguring materials, even some forms of distributed computing—keep returning to social insects for ideas. The ants are not offering a blueprint. They are offering a principle: let local load recruit local support, and let the support be made of the same units that do the moving.

In a well-formed living bridge there is a moment of visual clarity: carriers run over a chain of motionless bodies as if over a plank, barely breaking stride. The bridge ants' antennae wave, their legs grip, their abdomens press into one another. The surface is uneven at the scale of an ant, yet it is smoother than the jagged void it replaces. Overhead, the forest does not notice. Underfoot, the colony has quietly rewritten the terrain, not by changing the world permanently, but by briefly becoming the part of the world it needed.

5.3 The Wisdom of the Brake

A human traffic jam has a particular cruelty: nothing has to be wrong for everything to stop. No crash is required. No broken bridge. A few drivers tap their brakes, the gap behind them compresses, and the compression travels backward like a pulse through the line. From the air the highway looks calm. From inside a car it feels personal, as if the vehicle ahead is *choosing* to waste your evening. The truth is harsher and more interesting. Congestion is a collective pattern built from perfectly ordinary reactions.

Ants should be vulnerable to the same fate. Their trails can be crowded enough to look like living ropes, dense with outbound foragers and inbound carriers. They amplify good routes through positive feedback, which encourages yet more traffic onto the same lines. They sometimes build infrastructure that increases flow, which can concentrate traffic further. Everything about the system seems designed to produce runaway success until it collapses under its own weight.

And yet, for many species, the collapse is curiously hard to provoke. Ant trails can saturate without freezing. The secret is that acceleration is only half the story. The other half is braking: negative feedback mechanisms that turn crowding into a signal to slow recruitment, redistribute effort, and preserve flexibility.

Negative feedback is the principle behind most stable machines. A thermostat does not heat a room forever; it measures the temperature and reduces heating as the target is approached. Your body does not keep raising its temperature as you run; sweating is a feedback loop that counteracts heat. Without negative feedback, positive feedback is a fire. With negative feedback, it becomes a tool.

For an ant colony, the challenge is that no one can see the whole trail. There is no central meter of "crowding" and no committee deciding to close the on-ramp. Regulation has to happen locally, in the same spirit as trail formation itself. That means turning small, immediate experiences—a bump, a delay, a difficulty unloading food—into changes in signaling.

The simplest kind of brake is to stop laying down the accelerator.

In *Lasius niger*, a common garden ant that has become a workhorse of laboratory studies, crowding can reduce how much pheromone foragers deposit. Czaczkes and Grüter (2013) showed that as trails become crowded, fewer ants lay pheromone at all, and the reduction can be dramatic: across their conditions, the fraction of pheromone-layers dropped by roughly a factor of 5.6 from least to most crowded. Importantly, the ants were not responding to a chemical inhibitor painted onto the trail; they were responding to social density itself, assessed through encounter rate. In clever manipulations, objects coated with nestmate cuticular hydrocarbons—chemical signatures of "one of us"—could mimic some of the effect: bump into something that smells like a nestmate often enough, and you behave as if the trail is crowded.

That is negative feedback in a particularly elegant form, because it attacks the very substance of positive feedback. The colony does not need to invent a separate "anti-pheromone" message. It can simply modulate the production of the pro-pheromone signal. When the trail is empty, pheromone is cheap and useful; each deposit has a high marginal value because it helps recruit more workers to exploit a resource. When the trail is already packed, additional recruitment is less valuable and may even be harmful. Under those conditions, reducing pheromone-laying is a way of saying, in chemical terms, *enough*.

It also provides a route to flexibility that does not rely on waiting for pheromone to evaporate. Pheromone decay is a slow kind of forgetting, governed by chemistry and weather. If a colony had to depend only on decay to change its mind, it would be trapped by its own history. Grüter and Schürch (2012) emphasized that crowding-based negative feedback can allow colonies to avoid getting locked into a choice simply because they were early. When two options are available and conditions change—one food source becomes better, safer, or easier to exploit—the colony can reallocate foragers quickly, even if a strong trail already exists. In this picture, the "memory" in the environment is not erased, but it is overridden by local signals that tell ants the current state of the system is saturated.

This begins to sound like the behavior of a well-run restaurant.

At the beginning of the evening, an empty dining room is a problem. Staff might stand near the door, menus visible, smiles ready, because the marginal benefit of recruiting another table is enormous. Later, when every table is full and the kitchen is overwhelmed, the same restaurant becomes selective: reservations only, a waiting list, a host slowing the intake to match capacity. The restaurant does not become less hospitable; it becomes less willing to amplify demand it cannot serve. Ant colonies face an analogous constraint in foraging: food must be transported, processed, and shared. Recruitment that exceeds handling capacity produces delays that ripple outward.

The ant version of a waiting list is often tactile rather than verbal. An ant that experiences repeated head-on collisions can slow, turn aside, or stop. Those micro-stops can widen the effective trail by creating lanes; in some species, two-way traffic self-organizes into directional lanes, reducing conflict. But lane formation is not always sufficient. At very high density, any movement can become difficult, and the system needs a deeper brake than mere courtesy.

One provocative possibility comes from Argentine ants, a species notorious for forming massive invasive supercolonies in many parts of the world. In dense Argentine ant trails, researchers have reported that a minority of individuals can remain stopped for long periods, acting as obstacles. At first that sounds like a failure: why would you place bollards in the middle of a highway? Yet the counterintuitive suggestion is that these stopped individuals can stabilize flow by preventing the trail from collapsing into a chaotic scrum. Obstacles can force moving ants into more ordered lanes, reducing cross-traffic and allowing the stream to maintain a plateau of high throughput rather than suffering a sudden drop to near-zero. The idea resembles a trick used in crowd management: strategically placed barriers can prevent dangerous compressions and keep pedestrian flow smooth by channeling motion.

The image is uncomfortable because it again involves bodies used as infrastructure, but the function is different from the living bridges of army ants. The stopped Argentine ants are not primarily shortening a path across a gap; they are introducing friction and structure into an already overexcited system. The brake is not only chemical; it is physical.

Human engineers have long exploited the same principle. The first traffic circles and roundabouts, developed and refined in Europe in the early twentieth century, were designed to reduce catastrophic stop-and-go patterns at intersections by replacing them with continuous, regulated merging. Even more striking are "speed humps" and chicanes in residential streets: deliberate obstacles introduced not because they make travel faster but because they prevent the system from entering a dangerous regime. A perfectly straight road invites speed; speed increases the severity of mistakes; mistakes require hard braking and can create jams and crashes. A small, persistent inconvenience can be a net gain.

Ants seem to have learned that lesson by evolution. Under dense conditions, a colony may do better by accepting a lower maximum speed in exchange for avoiding catastrophic collapse.

All of these brakes raise a deeper question: how does an ant know when the system is nearing trouble? The answer is that trouble has local signatures. A bottleneck is felt as delay. High density is felt as frequent contact. Limited unloading capacity in the nest is felt as waiting with food in your mandibles while another ant fails to take it. A dangerous foraging site is felt as attack, alarm, injury, the sudden absence of returning nestmates.

In honey bees, negative feedback takes a form so explicit it can sound almost human: a "stop signal." Honey bee foragers recruit nestmates to food sources by performing the waggle dance, a symbolic movement that encodes direction and distance relative to the sun. The waggle dance is often held up as an example of insect sophistication, but sophistication brings a risk: a powerful recruitment signal can continue to drive foragers to a site that has become dangerous or unprofitable. Bees possess an inhibitory communication that can interrupt this recruitment. Stop signals are short vibrational pulses delivered by one bee to another, often directed at dancers, and they can reduce waggle dancing and recruitment. Researchers have linked stop signaling to situations where the colony's capacity is constrained—for example, when foragers cannot unload nectar quickly enough—and to contexts involving conflict or danger. Modeling work has suggested that stop signals can be adaptive during aggressive competition, rapidly shutting down recruitment to hazardous locations where continuing to send workers would be costly.

A bee delivering a stop signal is not being pessimistic. She is acting as a mobile brake pedal, transmitting a local assessment (*this is too costly right now*) into a change in global behavior (fewer recruits).

The comparison between ants and bees is illuminating because it shows two ways to implement negative feedback. Ants often regulate by altering a shared environmental signal: less pheromone, less recruitment. Bees regulate by direct inhibitory contact between individuals: a stop signal that interferes with the dancer's momentum. Both are local. Both are scalable. Neither requires an overseer.

They also show different relationships to memory. Pheromone trails are a kind of external memory: a record of past success written onto the substrate. Stop signals are more like real-time governance: they operate in the moment, independent of any chemical trace that may persist. In a fast-changing environment, governance can matter more than memory. If a predator begins attacking foragers at a site, the colony cannot afford to wait for the scent trail to fade. Inhibitory signals allow a rapid pivot.

The interplay between positive and negative feedback becomes clearest when the colony faces a choice between two options that are initially similar. Positive feedback can push the colony to commit to one. Commitment has benefits: it creates strong trails, efficient transport, and shared attention. But commitment has a hidden cost: it reduces exploration. The colony becomes less likely to notice changes elsewhere. Negative feedback counteracts that by keeping the system from saturating too completely. When crowding reduces pheromone deposition, a trail that is already busy stops becoming disproportionately busier, leaving some foragers free to sample alternatives. In other words, braking preserves a trickle of curiosity.

This is not mere metaphor. Consider a modern phenomenon: online recommendation systems. A video that receives early attention is promoted, which generates more attention, which generates more promotion. That is an engineered positive feedback loop. Without brakes, it can create monocultures of attention where a few items dominate for reasons unrelated to quality. Platforms therefore add brakes of various kinds: diversity constraints, downranking of repetitive content, friction in sharing. Whether these brakes are well designed is another question, but the logic is the same. A healthy information ecosystem needs both amplification and inhibition.

Ant colonies face an analogous design problem, solved not by software engineers but by selection acting over countless generations. A pheromone trail is an attention system. It directs workers toward a resource. If attention becomes too concentrated, the colony becomes brittle.

The most satisfying aspect of the ant brake is that it can be explained without imagining that ants are weighing pros and cons. The computation is performed by the pattern of interactions.

When a trail is lightly trafficked, an ant returning from food to nest experiences few collisions. She moves smoothly. Smooth movement correlates with high payoff to recruiting more workers, so pheromone-laying is likely. As the trail fills, collisions increase. Collisions slow movement and also directly cue ants to lay less pheromone. Less pheromone reduces the bias at the trail's entrance, so fewer new ants join. The system stabilizes around a density where traffic is high but not paralyzing.

This is why negative feedback deserves the label *wisdom.* It is not wisdom in the sense of foresight or moral judgment. It is wisdom in the engineering sense: a built-in tendency to keep the system within a workable range. A colony that lacked this tendency would be a colony that could discover a great food source and then fail to exploit it because its own success would choke its routes.

There is also a gentler emotional resonance here. Braking is a kind of restraint, and restraint is often treated as the opposite of intelligence: the smart move, we tell ourselves, is to seize the opportunity, to go all in. Ants show that restraint can be the smart move when the world is crowded and resources must be shared across many needs. The colony is not maximally fast; it is robustly fast. It does not win by sprinting until it collapses. It wins by staying in motion.

At dusk in a garden, you can sometimes see a thin file of ants moving along a crack in the pavement, each ant touching the ground with antennae as if reading braille. If you place a small crumb nearby, the file thickens. For a while it looks like triumph: a living road built in minutes. Then, just as quietly, it stops thickening. The trail remains busy but not frantic. A few ants peel off to explore the edges. The system has found a limit without being told there is one, guided by nothing more than the pressure of bodies and the ebb and flow of signals that those bodies choose to emit.

Chapter 6

Builders Without Blueprints

Imagine a skyscraper built by millions of blind workers who possess no blueprints and recognize no foreman, yet achieve precise climate control. This chapter demonstrates how complex architecture emerges from simple local interactions, transforming the termite mound from a pile of dirt into a breathing, self-regulating physiological extension of the colony itself.

6.1 The Algorithm of Mud

A termite carries a pellet of damp soil in its jaws the way a mason carries a brick: with purpose, weight, and a certain inevitability. Seen one at a time, the act looks almost trivial. The pellet is set down, pressed into place, and the worker turns away to fetch another. Yet out of countless such gestures arise structures that look engineered: buttresses and chimneys, vaulted passages, ribs and flutes. The puzzle is not that termites can move mud. The puzzle is how they move it *together*.

For a long time, the tempting answer was the human one: there must be instructions. Surely some termite, or some caste, must know the plan. Surely there must be a signal that says, *pillar here*, *wall there*, *arch now*. The real answer is stranger and, in a way, more elegant. The plan is not carried in any head. It is carried in the work itself.

In 1959, the French zoologist Pierre-Paul Grassé gave a name to this kind of coordination: *stigmergy*. The word was meant to capture a simple but radical idea. Individuals do not need to tell one another what to do, and they do not need an internal picture of the final structure. Instead, each worker responds to traces left in the environment by other workers. The mound becomes a medium of communication, and the building process becomes a loop: work produces a cue; the cue triggers more work; the new work strengthens or modifies the cue. A colony, in effect, thinks with mud.

Stigmergy is easy to appreciate if one begins with the limits of a termite. A termite is not blind in every sense, but it does not see like we do. It lives in a world of close-range textures, chemical gradients, drafts of air, minute changes in humidity, and the feel of a surface under its antennae. Its decisions are local, made at the scale of millimeters. A termite cannot stand back, squint, and judge whether a pillar is straight. It cannot walk a hundred body-lengths to check alignment with a distant wall. And yet alignment happens. Symmetry happens. Repair happens. The architecture is coherent. The coherence, the trick of it, is not imposed from above; it emerges from many small acts that tug in the same direction because they are pulled by the same kinds of cues.

The simplest version of stigmergy can be told as a story about a pile.

Imagine a patch of ground where a colony has begun to deposit pellets of wet soil. At first, the placement is close to random. A termite arrives with a pellet and drops it somewhere along a rough boundary. Now add one more ingredient: the pellet is laced with a chemical trace, a short-lived scent that other termites can detect. That trace does not need to be a sophisticated message. It can be nothing more than a label that says, *recent building happened here.* When another termite arrives carrying its own pellet, it does not choose a site uniformly. It is slightly more likely to drop the pellet where the scent is stronger, because that indicates prior deposition. A tiny bias replaces pure randomness.

This is the probabilistic rule that makes the world change: *place your pellet where the cue is a little stronger.* Not always. Not with certainty. Just with higher probability.

From the outside, that bias looks almost too weak to matter. But a weak preference, repeated thousands of times, becomes a powerful sculptor. A spot that happens, by chance, to receive two pellets instead of one becomes a slightly stronger source of scent. That spot is then a slightly more attractive target for the next passer-by. The next pellet makes the spot stronger still. Very quickly, what began as statistical noise becomes a seed: a little heap. The heap draws more pellets, rising into a nub, then a small column. The column has no architect. It has only a feedback loop.

Positive feedback is the engine here: success breeds more success. The termites are not cooperating by exchanging messages; they are cooperating by making the world uneven in a way that steers the next action. The mound is not merely built; it is also *read.*

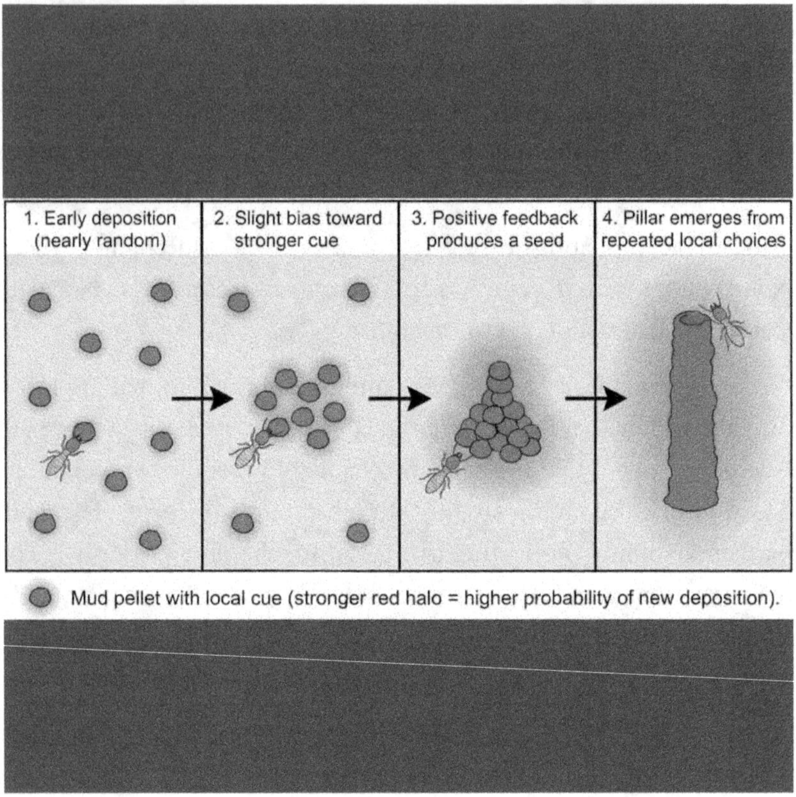

Stigmergic accumulation: a weak preference for already-marked sites turns chance clumps into stable pillars.

Once pillars appear, the story becomes richer, because the same local rule can generate more than one kind of global form. A pillar changes the environment. It reshapes airflow in a tunnel. It casts a tiny region of shelter. It presents curvature, edges, and a vertical surface to climb and sense. Even if the chemical trace fades quickly, the physical trace persists. The structure itself becomes a long-term memory.

This is one reason stigmergy is not just "pheromones did it." Pheromones are real, powerful, and widely used; but the broader point is that termites can recruit *any* reliable cue that correlates with successful building. A deposit of mud can carry a chemical tag; a curved surface can bias where more mud is placed; a humid boundary can indicate "inside" versus "outside." The colony does not need to agree on the meaning of the cue in a linguistic sense. Natural selection has already done the agreeing by shaping what a worker tends to do when it encounters that cue.

To see why this matters, consider what a blueprint really is. In human construction, a blueprint stores information about a building before the building exists. It is a map of the future. Termites have no such external representation. But they do not need one if the partially built mound can serve as a continuously updated instruction sheet. The instructions are not written in symbols. They are written in gradients and shapes and resistances: in where the surface yields under the mandibles, in where the air feels drier, in where traffic is easier, in where the material already holds. A human foreman tells you to add bricks to make a wall. A mound tells a termite where a wall is likely to become useful simply by being the kind of place that attracts another pellet.

It helps to be concrete about what "attracts" means in practice. In many termites, building material is not placed gently; it is tamped, smeared, and fused. Pellets are made of soil and saliva, sometimes enriched with fecal matter, and can include chemical compounds that evaporate or diffuse. If a fresh pellet releases an odor that fades over minutes, then a spot with many fresh pellets is, effectively, a high-intensity beacon. That intensity can be read as a scalar: more smell, more likely deposition. Eric Bonabeau and colleagues, in a series of models in the late 1990s, showed how far such a minimal assumption can be pushed. With agents following simple rules-wander, pick up a pellet, drop it preferentially where a virtual pheromone is high-pillars appear reliably. Under slightly different boundary conditions, pillars can link into walls. With additional templates, galleries and chambers emerge. The striking part is not the exact architecture of any one simulation. It is the idea that the *same* behavioral toolkit can yield different outcomes when the environment is different, because the environment is not a passive stage. It is a participant.

That last point can sound abstract until one pictures a worker inside a mound. The worker is constantly in motion, surrounded by other workers, in a corridor that is never finished. There is no moment when "construction" stops and "use" begins. The mound is used while it is built; it is built while it is used. That entanglement means that cues are everywhere. The flow of bodies becomes a cue. The ease of passing becomes a cue. The accumulation of carbon dioxide becomes a cue. A local traffic jam is not just an inconvenience; it is information about where space might need to expand.

Stigmergy also dissolves a common misunderstanding about collective intelligence. One imagines that for a group to be smart, individuals must have a shared plan and then coordinate their roles. Termites do coordinate roles, of course-there are castes, and tasks are not distributed at random-but stigmergy shows that coordination can be achieved by something even simpler: *compatibility* between a behavior and the traces it tends to create. If dropping pellets where pellets already are produces sturdy structures, then a colony in which workers have that bias will, over time, build sturdier structures. The "intelligence" is in the fit between rule and world.

There is an old-fashioned naturalist's pleasure in imagining how Grassé came to his idea. One can picture a laboratory observation that refuses to behave like a tidy hierarchy. You remove a piece of a nest, and instead of waiting for a "boss" to arrive and direct repairs, workers begin immediately, in parallel, wherever the disturbance is strongest. Plugging begins where the air rushes. Deposition concentrates where the edge is raw. The activity does not look like obedience; it looks like attraction. It looks like the work itself is calling for more work.

The historical anecdote matters because it reminds us that stigmergy was not coined to be fashionable. It was coined because a careful observer needed language for a phenomenon that did not fit the command-and-control instincts of human explanation. Once you have the language, you begin to see stigmergy everywhere. Ant trails are the familiar example: a few ants lay a weak chemical track; others follow it; the track strengthens; soon a stable path appears. But termite building is a deeper demonstration because the "trail" is not just guidance through space. It is a transformation of space. The cue is not merely a line to follow; it is a scaffold that becomes the structure.

There is a further twist. Positive feedback can run away. If termites only amplified existing deposits, they might end up with pathological spikes-too many pellets on one pillar, too little elsewhere. Real mounds are not just exuberant; they are restrained. That restraint comes from negative feedback: processes that dampen growth and prevent lock-in to a bad early accident. Some negative feedback is mundane. The chemical cue fades, so an abandoned site stops attracting attention. Some is mechanical. A pillar that becomes too tall may be unstable; pellets may not adhere as well; workers may be less likely to climb it if the route is awkward. Some is social in the simplest sense: there are only so many workers, and time spent reinforcing one site is time not spent elsewhere. Stigmergy does not deny these constraints; it relies on them. A mound is a negotiation between amplification and limitation, between the urge to pile and the need to keep passages open, between growth and use.

That negotiation also clarifies why "simple rule" does not mean "simple outcome." The rule is simple in the sense that each act is local and does not require a global map. But the system of cues that a termite responds to is not one-dimensional. Chemical traces, airflow, humidity, surface curvature, and tactile texture can all be part of the same decision. Experiments in the last decade have underlined how strongly physical geometry can guide early deposition, as if the curve of a surface were itself a sign. It is not that pheromones vanish from the story; it is that pheromones may be only one channel among several. Mud is not merely a building block. It is a sensory surface.

Here the "algorithm" metaphor earns its keep. An algorithm is not a plan that contains a finished picture. It is a procedure that transforms an input into an output by repeated local steps. Stigmergy turns the construction site into a kind of external memory: each deposited pellet updates the state of the site, and the updated site changes the probabilities of the next deposition. In computer science, one might call that a distributed computation. In ordinary language, it is the feeling of a place that begins to "want" to become something. The mound, once seeded, pulls itself into being.

This way of building has a curious emotional resonance for humans, because it suggests a kind of creativity without imagination. The termites do not daydream about arches. They do not admire symmetry. They do not argue over aesthetics. And yet the results can look, to our eyes, aesthetic: smooth buttresses, repeating ribs, a suggestion of intention. The intention is real, but it is not located where we instinctively look for it. It is located in a feedback loop stabilized over evolutionary time.

There is also a practical, almost philosophical implication. When the environment stores the instructions, the system becomes robust to the failure of any individual. Lose a worker, and the building does not halt. Lose a hundred, and the building slows, but it does not forget what to do, because what to do is written in the partially built structure and in the cues it generates. This is a kind of resilience that centralized systems envy. A human building site without its foreman is a different place. A termite mound without any particular termite is still, in a meaningful sense, the same project.

Stand close to a mound and imagine the rule, repeated: carry, deposit, press; carry, deposit, press. Imagine the slight bias toward what is already there. Imagine the world becoming uneven, and the unevenness becoming a guide. Somewhere in that loop lies the secret that makes a heap become a pillar, and a pillar become a wall, and a wall become a home that can last for years. Not because anyone knows the whole, but because each tiny act, guided by the traces of other tiny acts, makes the next act more likely to fit. The mud does not just hold the mound up. It tells the colony where the mound wants to grow.

6.2 The Living Lung

A mound can look like a monument to endurance: sun-baked earth shaped into towers and ridges that survive storms, hoofbeats, and years of weather. From the outside it seems inert, almost geological. Yet the colony inside treats it less like a castle and more like a bodily organ. The mound is not simply where termites *are*; it is something they *use* to stay alive.

The trouble begins with an awkward fact about termite life. A dense society, underground, breathing and producing waste gases, is at constant risk of suffocation. Every termite consumes oxygen and releases carbon dioxide. In many of the most spectacular mound-builders, there is a further burden: a fungal agriculture. Fungus-growing termites cultivate fungus in subterranean combs, and the fungus has its own respiratory needs. Add heat produced by metabolism, water that must not be lost too quickly, and the challenges of being small-bodied in a harsh climate, and the nest becomes a balancing act between competing demands. Open the system too much and it dries out, cools too far at night, and becomes vulnerable.

Seal it too tightly and carbon dioxide builds up, oxygen drops, and heat and humidity can drift out of the narrow range that sustains the colony and its crop.

The mound is the colony's answer to this problem: a porous, sculpted interface between an inner world and the atmosphere. It works without fans, valves, or centralized control. It works because the physical structure is shaped so that ordinary forces—wind, sunlight, and daily temperature swings—do useful work. The mound "breathes" not by muscular effort but by turning the environment into machinery.

To appreciate how strange this is, it helps to imagine a human building designed with the same constraint: no moving parts, no electricity, and no caretaker who understands the whole system. The building must nevertheless keep carbon dioxide low, oxygen high, and temperature and humidity within bounds, while its internal occupancy fluctuates and the weather outside changes by the hour. Human architects do attempt such feats, and sometimes borrow from termite mounds as inspiration, but the termite version has an additional twist. The architecture is not planned in advance; it is the emergent result of innumerable local decisions. The structure is both a product of swarm behavior and a tool that shapes that behavior.

A useful starting point is a simple physical idea: air moves when pressure differs. Pressure differs when air is heated, cooled, pushed by wind, or displaced by convection. Termite mounds exploit all of these, but not always in the same way. A temptation in popular retellings is to present a single, elegant mechanism and declare it *the* solution. The science is more interesting than that. Different mound shapes, different habitats, and different termite species can yield different ventilation regimes. A mechanism that dominates in a sunlit savanna can be muted in a shaded forest. A mound that is tall and fluted will interact with wind and sunlight differently from one that is low and ridged. The "lung" is not one design but a family of designs, each making trade-offs with the local climate.

The historical arc of this research has its own drama. For decades, there were canonical explanations for mound ventilation that sounded plausible and were repeated often. They had the tidy feel of textbook diagrams: warm air rises up a central chimney; cooler air is drawn in from below; the mound functions like a stove pipe. Such stories appealed to common sense, and it is easy to see why they took hold. But termites do not owe us intuitive mechanisms. When Judith Korb and Klaus Linsenmair reported field measurements in 2000 that challenged widely accepted ventilation hypotheses, the lesson was not merely that a specific model was wrong. The deeper lesson was methodological: mounds are complicated physical objects embedded in complicated environments, and a story that fits on a napkin can fail in the field.

Fieldwork is humbling because a mound is not a laboratory apparatus. The airflow is subtle and can reverse with time of day. Wind can change by the minute. The mound's outer walls are not uniform; they are porous in one place and dense in another, wet after rain and brittle after drought. The internal channels are not smooth pipes but rough, branching passages crowded with insects. Measuring flow in such a system requires patience, cleverness, and the willingness to accept messy results.

In the last decade, a particularly compelling picture has emerged for some large mounds: ventilation driven by *diurnal temperature oscillations*—the daily warming and cooling cycle imposed by the sun. Hunter King, Samuel Ocko and colleagues reported in 2015 that some Macrotermes mounds harness these oscillations to drive a circulating airflow that flushes respiratory gases. The key is that different parts of the mound change temperature at different rates. Thin outer structures, exposed to sunlight and nighttime cooling, heat and cool relatively quickly. Thick central regions, shaded and thermally massive, change more slowly. When one region is warmer than another, buoyancy differences arise: warmer air tends to rise relative to cooler air. In a suitably connected network of channels, that buoyancy can produce a circulating loop, like a convection cell.

A striking feature of this mechanism is that the direction of circulation can reverse between day and night. When the outer walls warm during the day, they can drive one pattern of flow; when they cool at night, the gradients invert and the circulation flips. The mound is not a chimney with a single one-way draft. It is more like a breathing pattern with a rhythm set by the sun.

This is where the mound begins to feel organ-like in a very literal sense. A lung is not simply an empty bag. Its structure creates conditions for exchange. The termite mound's network of pores and channels increases surface area for gas diffusion, while convection and wind-driven pressure differences help move air through that network. The colony's metabolism supplies carbon dioxide that must be removed and heat that must be managed; the mound shapes the flows that accomplish that removal and management. The "patient" and the "device" are not separable. Without termites, the mound is a dead husk. Without the mound, the termites cannot easily sustain their dense underground life.

It is tempting to picture ventilation as serving only respiration, as if the mound's job were simply to keep oxygen coming and carbon dioxide going. But in many termite systems, the microclimate is just as important. The fungus gardens of fungus-growing termites, for instance, have their own requirements for temperature and humidity. Even termites that do not farm fungus are sensitive to desiccation. A mound that ventilates too aggressively could turn into a dehydrating wind tunnel. The architecture must allow exchange without sacrificing stability.

This is where the porous outer wall becomes crucial. A mound's skin is not a sealed shell; it is a regulated interface. Porosity can vary with height and location, and it can change as the termites plaster and repair. A porous wall allows diffusion: gases can seep through without needing a dramatic breeze in a single pipe. A channel network allows convection: bulk movement of air that can flush regions more quickly than diffusion alone. Wind can add another layer, creating pressure differences around the mound's exterior that drive flow through openings and pores. In many cases, the mound's external ridges and flutes may help generate such pressure differences, turning passing air into a pump. But the relative importance of wind, metabolic heating, and solar-driven temperature gradients can differ between systems.

This diversity is not an embarrassment; it is the signature of adaptation. A mound is built by a colony embedded in a particular climate. The mound grows over years. Each season, the colony experiences the consequences of its structure: how quickly it loses water, how well it sheds excess heat, how effectively it maintains gas exchange. Colonies that build structures ill-suited to their environment do not flourish. Over evolutionary time, what persists are not the most *beautiful* mounds but the ones that maintain a livable internal atmosphere.

The beauty, though, is that the colony is not measuring "livable internal atmosphere" with instruments. Individual termites respond to local cues: humidity at the surface, carbon dioxide concentration along a tunnel, the feel of a draft through a breach, the temperature of a wall. Those cues feed into stigmergic building rules. The mound changes, the cues change, and the building changes again. The result can look purposeful even though it is built from local responses.

Consider how one might engineer a stable internal climate without understanding thermodynamics. The termites do not need to compute a convection equation. They need only a set of tendencies shaped by natural selection: deposit more material where air is too dry; open pores where carbon dioxide is high; reinforce walls where a draft is felt; build ridges where wind scours; avoid exposing fungus combs to abrupt temperature swings. Each tendency is local and immediate. The combined effect, across thousands of workers, is a structure that couples the nest to the outside world in a controlled way.

One can feel the drama of this coupling most vividly at the boundary between inside and outside. Inside, air can be humid and warm, heavy with carbon dioxide. Outside, air can be dry, cooler, and rich in oxygen. The boundary is where exchange happens, and exchange is always double-edged. Every molecule of carbon dioxide that leaves is a success, but every bit of water vapor that leaves may be a cost. Every influx of oxygen is welcome, but a sudden blast of cold air may be dangerous. The mound's function is not merely to maximize exchange, but to keep the exchange in a safe envelope.

This is why it is helpful to think of the mound as an extended physiology rather than a mere shelter. Human physiology does not simply let blood pressure rise as high as possible; it regulates pressure. It does not simply maximize sweating; it uses sweating in a context-dependent way. Termite colonies achieve regulation through architecture and behavior rather than through organs and nerves, but the end result is comparable: homeostasis at the level of the collective.

A concrete example makes this less abstract. Picture a hot day on the savanna. The mound's outer surfaces heat unevenly as the sun moves. Some surfaces warm directly; others remain shaded. The air in outer conduits warms and becomes buoyant relative to cooler air in deeper channels. A pressure gradient forms, and air begins to circulate through the channel network. Carbon dioxide produced by termites and their fungus combs is carried along and can diffuse out through porous walls or be vented through openings. As evening comes, the outer walls cool quickly while deeper regions stay warmer.

The buoyancy pattern inverts, and the circulation reverses. Over a full day-night cycle, the mound has used environmental oscillations as a pump. No worker has to coordinate the timing. The sun does it for free.

There is a kind of generosity in this arrangement: the mound is not fighting the environment; it is harvesting it. Sunlight, which can be dangerous to a small insect, becomes an engine for ventilation. Nighttime cooling, which can be dangerous for maintaining warm, humid conditions, becomes part of a cyclic flow that renews the air. Wind, which might strip moisture away, becomes a pressure field that can be exploited by ridges and pores. The mound turns the outside world's violence into internal regularity.

This idea also helps explain why measuring mound ventilation has been so contentious. If the flow reverses, if it is subtle, if it depends on sunlight and wind and the thermal properties of different mound regions, then a measurement taken at one place and one time can mislead. A researcher who arrives at midday might infer one mechanism; another who measures at night might infer the opposite. A researcher working in a different habitat might see a system in which wind plays a larger role than thermal oscillations. The correct picture is not a single diagram but a set of physical principles applied to a living, changing structure.

The mound's "lung" function has another consequence that is easy to miss if one thinks only about air: it shapes what it means for the colony to be a swarm. A swarm, in this context, is not merely many bodies in proximity. It is many bodies coupled through shared physical fields. Airflow is a field. Temperature is a field. Humidity is a field. Carbon dioxide concentration is a field. These fields are not abstract; they are felt by termites through sensors on their bodies. When the mound shapes the fields, it shapes the information available to individuals. It creates gradients that guide behavior: where to deposit mud, where to open passages, where to move brood, where to tend fungus. Architecture becomes both physiology and information-processing.

That intertwining can produce a pleasing paradox. The mound is built by termites responding to local cues, yet once built it becomes a device that generates and stabilizes those cues. It is a feedback loop on the scale of meters and years rather than millimeters and minutes. A worker may never experience the full long-term arc of mound growth, but its moment-to-moment choices are still shaped by the conditions that the mound creates. The colony's intelligence is, in a sense, offloaded into the wall thickness, the porosity, the channel geometry, and the thermal inertia of packed earth.

If this begins to sound like metaphor, it can be grounded again in the simplest physical picture: the mound is a heat-and-mass exchanger. Warmth moves through it. Water vapor moves through it. Gases move through it. Its shape and material properties determine how fast those flows occur and how strongly they vary over time. Termites modify those properties by adding and subtracting mud. The colony is continually tuning the parameters of its own life-support system.

Standing near a mound at dusk, one can imagine the air inside beginning to shift as the sun drops. The outer walls cool; the gradients change; the slow circulation rearranges itself. Deep underground, where the fungus combs sit like dark loaves in humid chambers, the atmosphere is being renewed by forces that have no intention but have consequences. The mound does not merely contain the colony. It mediates between a fragile living mass and an indifferent sky, breathing by shape alone.

6.3 The Self-Healing Fortress

A fortress, in the human imagination, is defined by its walls. A wall is a boundary drawn in stone: inside safe, outside dangerous. The termite mound also draws a boundary, but it draws it in a far more precarious medium—earth mixed with saliva, plastered into place grain by grain. The boundary is constantly challenged by weather, trampling animals, rival colonies, and predators that treat the mound as a pantry. What makes the termite system remarkable is not that it resists damage. It is that damage becomes part of how it functions. A breach is not merely a failure; it is a signal.

Human repair usually requires diagnosis. Someone notices a crack, decides it matters, gathers tools, assigns workers, and coordinates them. Termites repair without diagnosis in that sense. No individual termite needs to understand what has happened to the mound as a whole. A hole in the wall changes the local world: air moves differently, humidity drops, light penetrates, vibrations carry in unfamiliar ways, and foreign odors arrive. Those changes are exactly what termites can sense. The mound advertises its own injuries in the language the builders already speak.

The most intuitive injury signal is airflow. A termite nest is tuned, by architecture and by constant maintenance, to keep certain patterns of air exchange and certain pockets of humidity. When the wall is intact, a tunnel feels like a tunnel: still, dark, and buffered. When the wall is breached, the tunnel suddenly feels like the outdoors. A draft appears. The air becomes drier. Carbon dioxide gradients shift. The quiet pressure field that the mound uses to "breathe" is interrupted by a sudden shortcut. The injury is, so to speak, tactile.

A second signal is light. Many mound-building termites spend their working lives in darkness. Light entering a tunnel is not just illumination; it is an alarm. It correlates strongly with exposure and predation.

The colony does not need a concept of vulnerability. It needs only an evolved tendency: when bright light touches a normally dark surface, deposit soil and close it.

A third signal is humidity. The interior of an active mound tends to be moist relative to the outside air, even when the surrounding landscape is dry. That moist interior is not an accident; it is a product of the colony's metabolism, of water held in soil, and often of fungal agriculture that thrives in high humidity. A breach punctures what can be imagined as a "humidity bubble." Local humidity plummets near the opening, and that drop can bias building behavior. Work that might normally extend a tunnel outward can switch to plugging and thickening at the edge, because the edge now feels like a dangerous boundary rather than a stable frontier.

Repair is therefore not a special program held in reserve. It is a reweighting of ordinary building rules under extraordinary conditions. The same termite that deposits mud to thicken a gallery or smooth a wall can deposit mud to seal a breach; what differs is not the action but the urgency, density, and location of deposition. In a decentralized system, urgency does not need to be broadcast. It is physically present.

A vivid way to imagine this is to picture the mound as a distributed sensor network. Each termite is a mobile sensor with a few channels: chemical cues, moisture, airflow, contact, light. The mound itself is a scaffold that shapes these signals, and when the scaffold is damaged the signals change abruptly. The "network" does not send a message to a headquarters; it reshapes the statistical landscape of local stimuli, and termites respond where the stimuli are strongest. The colony's labor flows toward the wound in the same way water flows toward a depression: not by command, but by gradient.

Concrete observations bring this to life. In work on the termite *Odon-*

totermes formosanus, reported in 2018, researchers described how disturbance can provoke rapid switching between different behaviors. When an opening appears, termites may begin repair almost immediately, but their tactics are sensitive to context. If the disturbance suggests predator presence, sealing can become urgent, with termites capping openings quickly rather than building elaborate reconstructions. Under other conditions, they may rebuild tubes and passages in a more expansive way, restoring traffic flow while gradually re-establishing the protective envelope. That flexibility can look like strategy, but it can arise from simple conditional rules: light and draft trigger one kind of deposition; certain vibrations or chemical cues trigger another; congestion alters yet another.

This is the self-healing fortress at its most impressive: a wall that recruits its own masons. A breach creates a draft; the draft draws workers; workers close the breach; the draft diminishes; the stimulus fades; workers disperse. The repair effort ends not because a foreman declares the job done, but because the world stops asking for repair. The system has a built-in stop condition: the disappearance of the cue that triggered it.

There is a pleasing symmetry here with the way stigmergic building begins. A small accidental pile of pellets, marked by a trace, attracts more pellets until a pillar appears. A small accidental hole, marked by airflow and dryness, attracts pellets until a wall reappears. In one case the positive feedback amplifies construction; in the other, it amplifies restoration. Both are forms of local reinforcement around a salient feature. The difference is the sign of the feature: a bump versus a gap. A bump announces "building is happening here"; a gap announces "building is needed here."

Repair also exposes a subtlety that is easy to miss if one imagines termites as tiny robots following a single line of code. A robot might seal a breach by repeatedly dumping material into the opening until it is full. Termites must do something more delicate. A nest wall is not a plug in a bottle; it is part of a ventilating, porous interface. Seal too perfectly and the nest may become stifled. Seal too loosely and the breach remains a vulnerability. Termites often construct with a kind of layered pragmatism: immediate emergency closure first, then reinforcement, then reshaping and smoothing until the repaired region matches the mechanical and permeability properties of the surrounding wall. This can happen without a plan if different cues dominate at different times. At first, the draft and light are intense, encouraging fast deposition. Once the opening is partially closed, light drops and airflow becomes weaker; the system shifts toward finishing behaviors that are more typical of normal maintenance.

A historical anecdote from the broader study of social insects illustrates why this matters. Early observers, confronted with the speed and apparent coherence of insect repair, often reached for metaphors of discipline and command. The colony was likened to an army, a factory, a state. Such metaphors flatter human institutions by finding them in nature, but they also mislead. An army repairs by orders. A factory repairs by scheduled maintenance. Termites repair by physics. The wall does not wait for permission to matter; it matters immediately because the physical fields around it change.

That physicality has another consequence: the response can be graded rather than all-or-nothing. A small crack that produces only a slight draft may trigger occasional deposition and reinforcement. A large breach that floods a tunnel with moving air may trigger a rush of workers and intense activity. The colony does not need to classify the damage as "minor" or "major." The magnitude of the cue effectively encodes the magnitude of the problem.

This kind of graded response is one reason decentralized systems are often fault-tolerant. A centralized system can fail catastrophically if its control center is disrupted or misinformed. A decentralized system can absorb local shocks because control is not concentrated. Termites do not need to find the one termite who knows how to repair. Any worker that encounters the right stimulus can contribute. Even if many workers are lost, the rules still work for those that remain, and the strongest stimuli still recruit the most effort.

The same logic extends to the problem of congestion. Repair sites attract many individuals. If everyone rushes to the breach, tunnels could clog, and the very attempt to fix the problem could create new problems. Termites seem to manage this through a combination of physical constraints and behavioral flexibility. Narrow passages limit flow; workers may switch tasks when crowded; some individuals may carry soil in while others carry excavated material out; some may retreat when cues diminish. Again, no one needs to oversee traffic. The geometry of the tunnels and the short-range interactions between bodies can regulate throughput.

Humidity offers a particularly evocative lens on this. In some mound systems, the interior atmosphere extends slightly into the outside world at an active building edge: a small halo of humid air just beyond the wall. That halo can act as a template for where building is feasible or desirable. When wind or dry air strips the halo away, the "outside" effectively moves inward, and deposition shifts toward sealing rather than expansion. A breach is, among other things, a violent disruption of this atmospheric boundary. The repair that follows can be understood as an attempt to restore not merely the wall's shape but the wall's capacity to maintain an inner climate. The fortress heals in order to keep the lung functioning.

This has a human resonance because it recasts what it means to be "safe." For termites, safety is not simply the absence of predators. It is the maintenance of a particular internal world: dark, humid, chemically buffered. A breach is dangerous because it dissolves the boundary conditions that make that world possible. Repair is therefore less like patching a hole in a fence and more like sealing a leak in a pressurized habitat.

The materials used in repair are humble, but the logic is sophisticated. A worker brings a pellet of soil mixed with saliva, presses it into place, and the pellet becomes part of a composite wall. The wall is simultaneously structural and permeable. It must be strong enough to resist collapse and intrusion, yet porous enough to participate in gas exchange. A repair patch must blend into this dual-purpose material. If one watches repair over time, one can often see the shift from emergency sealing to architectural integration: rough plugging gives way to smoothing, thickening, and the restoration of a continuous surface that can once again serve as a reliable cue landscape for ordinary building.

There is also a quiet lesson here about the relationship between damage and information. In many human systems, damage destroys information. A broken sensor sends no signal; a collapsed database loses its records. In termite systems, damage *creates* information by generating gradients: of airflow, of light, of humidity. The injury broadcasts itself. This is why the mound can be called self-healing in a literal sense. It is not that the mud has magical regenerative powers. It is that the colony has evolved to treat certain physical disruptions as urgent stimuli, and those stimuli are inevitably produced by disruption.

One can imagine this as a kind of moral economy of attention. A well-functioning mound is, from the termite's perspective, boring: stable air, stable humidity, stable darkness. Boredom is productive because it allows workers to distribute themselves across routine tasks. A breach is exciting in exactly the wrong way. The world suddenly contains a sharp feature, and sharp features recruit behavior. As the feature is blunted—as the hole shrinks, the draft weakens, the light fades—the excitement disappears and attention diffuses again. The fortress heals by allowing attention to be captured and then released.

Standing near a mound after a storm, one might see a fresh scar: a patch where the texture is different, where the soil looks newer, where the surface bears the marks of recent plastering. Beneath that patch, the colony has already experienced the crisis and muted it. Airflow has been redirected back into the mound's slow breathing circuits. Humidity has been recontained. Darkness has returned. The repair is not celebrated, not recorded, not even remembered by any individual in a way that lasts. It persists as geometry and material, a quiet demonstration that a distributed mind can treat catastrophe as just another kind of cue, and answer it with mud.

Chapter 7

Decoding the Swarm

To the naked eye, a flock of starlings is a shapeshifting phantom; to the researcher, it is a massive dataset waiting to be parsed. This chapter dismantles the methodology of swarm science, tracing the journey from the limitations of manual field notes to the precision of high-speed computer vision and the digital alchemy of agent-based modeling.

7.1 The End of Telepathy

Stand at the edge of a winter roost when starlings pour into the dusk and it is hard not to feel that something *extra* is happening. A thousand bodies fold and unfurl as if a single muscle lay under the sky. A predator cuts through the margin and the whole cloud seems to tense at once, pinching inward, then spilling outward in a dark ribbon. To a human observer, the coordination looks instantaneous. If each bird had to take a vote, or wait for a signal from a leader, the flock ought to crumple into confusion. Yet it doesn't. For a long time, that gap between what the eye reports and what common sense expects invited supernatural language.

Victorian and early twentieth-century natural history was full of it. Popular writers reached for a "group spirit", a "collective mind", a "racial memory", a "sympathy" that travelled faster than sensation.

Even careful observers sometimes used words like "as if by telepathy" not as a claim to the paranormal, but as a confession of helplessness: they had no satisfying mechanism to put in its place. It is worth taking those mystical explanations seriously, not because they were correct, but because they were a reasonable response to an information problem. If you cannot measure who reacts to whom, if you cannot see the tiny delays between one animal's turn and another's, then the group behaves like a sealed box. When the only observable fact is the clean, global pattern, a global cause feels tempting.

Two things had to change for "telepathy" to die. The first was conceptual: scientists had to accept that complicated, purposeful-looking patterns can be produced by individuals following crude rules with no overview. The second was technical: they needed ways to *show* those rules at work, by tracking individuals accurately enough to see local interactions adding up to collective order. The story of how those changes arrived is not the tidy replacement of superstition by reason. It is a slow conversion, driven by arguments about what animals can plausibly sense, by new mathematics, and by the sheer stubbornness of data.

A useful starting point is to notice how the illusion is built. When a flock wheels, you see the entire flock wheel. Your brain compresses time; it treats the event as one action, one intention. But a flock is not a rigid object. It is a medium, more like a crowd in a train station: each person can only watch nearby bodies, and any "decision" that spreads through the crowd must travel, step by step, through those local observations. In the nineteenth century, the physics of such spreading processes was still young. The idea that a disturbance can propagate through a material without any one part "knowing" the whole—a wave through water, a pressure pulse through air—was available, but not yet a default metaphor for living groups. It did not help that the medium here is made of agents, each with eyes and nerves, so the wave is not purely mechanical. The coordination is *computed* by many small brains in parallel.

One of the earliest clues that mystical explanations were unnecessary came not from birds but from insects. Consider a line of ants finding a food source. To a casual observer the line looks planned: a straight road laid down in advance. Naturalists once suggested that ants must be guided by a communal intelligence, or that a scout somehow instructed the whole colony. In fact, the most powerful ingredient is almost embarrassingly simple: a chemical trace. An ant that stumbles upon food returns to the nest while laying pheromone; other ants are slightly more likely to follow the scented path; those that follow successfully reinforce it with more pheromone on the way back; and very quickly a thin bias becomes a highway. The "plan" is not in any individual. It is in the feedback loop between individual motion and a shared environment.

That ant trail is more than a charming example. It teaches a general lesson: a group can seem to make a decision when it is really amplifying tiny statistical differences. In human terms, it is like applause that suddenly synchronizes not because one person commands it, but because everyone is listening, adjusting, and copying what is already slightly dominant. Once you have that lesson in mind, telepathy begins to look like an unnecessary extra layer. The question shifts. Instead of asking, "What invisible force coordinates the group?", you ask, "What local signals are available, and what simple responses to those signals could produce the global pattern we see?"

That shift sounds obvious now because it matches a broader transformation in twentieth-century science. Biology, once dominated by description and classification, began to borrow the mathematical habits of physics: build minimal models, derive consequences, test them against measurements. The transformation had famous milestones—population genetics, enzyme kinetics, epidemiology—but collective behaviour posed a special challenge. The system is large (often thousands of individuals), fast (turns can unfold in seconds), and messy (animals obscure one another, and the group is never neatly bounded). In practice, group behaviour remained a realm where metaphor could survive longer than it should have, because the data needed to kill it were difficult to obtain.

When the modelling turn finally took hold, it did so by insisting on a certain kind of humility: assume individuals have limited perception; assume they respond only to nearby neighbours; assume there is no leader unless you can demonstrate one. In the 1990s, Tamás Vicsek and colleagues made that humility mathematically explicit in what became a foundational model of collective motion. Their agents were not clever. Each one moved at constant speed, repeatedly adjusted its direction to match the average direction of neighbours within a local neighbourhood, and suffered a little noise. That was all. Yet as the density increased or the noise decreased, a striking transition occurred: the group shifted from disordered milling to coherent motion in a common direction.

The importance of such a model is not that it is a literal description of any one species. It is that it showed, with the clarity of an equation, how "mind-like" group behaviour can emerge from rules that are mindless in the everyday sense. A flock does not need a plan to move as one. It needs alignment, locality, and a way to transmit influence through repeated neighbour-to-neighbour updates. In physics, one might call this an *order—disorder transition*. In dinner-party language, it is the moment a crowd stops being a collection of individuals and starts behaving like a single moving thing.

Even with models like this on the table, scepticism remained. Mathematical biology can be intoxicating; one can generate convincing patterns on a computer with rules that no animal would ever follow. The old mystery threatens to return in a new costume: instead of telepathy, you have a simulation that "just happens" to look right, without being tethered to what real animals actually sense and do. To finish the job of demystification, the field needed measurements that could discriminate between plausible local rules and merely pretty ones.

A concrete illustration of that discrimination came from an Italian collaboration led by researchers including Michele Ballerini and Andrea Cavagna, who in the late 2000s and early 2010s used multi-camera methods to reconstruct the three-dimensional positions of starlings in large natural flocks. The technical achievement alone mattered: it turned what had been a poetic spectacle into a dataset. But the conceptual result was sharper. By analysing how each bird's motion correlated with its neighbours, they argued that starlings interact not with all birds within a fixed metric radius (say, every neighbour within two metres), but with a fixed *number* of nearest neighbours, roughly six or seven, regardless of the absolute distance.

This is a small, almost domestic number. It punctures the telepathy illusion in a very specific way. If each bird is effectively "listening" to only half a dozen others, then no individual has access to the state of the flock. Any group-level coordination must be a property of the network formed by these local attentions. Yet the fixed-number rule is powerful because it is robust to changes in density. In a tight cluster, six neighbours are very close; in a stretched-out formation, six neighbours may be farther away. Either way, the bird maintains a workable sample of the local crowd. During a predator attack, when a flock can compress in one region and dilute in another within moments, a metric rule can fail: individuals in a sparse region may find themselves with too few neighbours inside the radius and "fall off" the group. A topological rule—fixed neighbour count—is less brittle.

Notice what kind of explanation this is. It is not a romantic claim that starlings share thoughts. It is a claim about *sampling*. It says: a bird's nervous system likely cannot track hundreds of individuals, but it can track a handful; it does not need distances in metres, just relative positions; it updates continuously; and the network built from these small attentions is enough to hold a flock together through violence and surprise. The mystery is translated into something that can be measured: how many neighbours matter, how fast individuals respond, how influence propagates.

That propagation is another place where telepathy used to hide. If the edge of a flock senses danger first, why does the centre turn so quickly? A naïve guess would be diffusion: a slow, lossy spreading of information like dye in water, where the signal weakens as it travels. But careful measurements of real collective turns in starling flocks suggested something more like a wave: a directional change can travel across the group with little attenuation, maintaining coherence over large distances. This matters because it shows that "instantaneous" coordination is often merely "fast enough to fool the eye," and it encourages a mechanistic vocabulary: propagation speed, delay times, correlations.

Correlation is a subtle word that deserves a plain meaning. If two birds are correlated, it does not mean one commands the other. It means that when one deviates from the average—turns slightly left, accelerates a little—the other tends to deviate in a related way. Correlations can extend far beyond the direct interaction range. You can have a system where each individual only looks at six neighbours, yet a change at one end influences the other end because influence chains through many intermediate links. A line of people passing a bucket works the same way. No one can reach the fire alone, but the bucket travels across the line quickly because each person only needs to coordinate with adjacent hands. The "group mind" becomes a property of the connectivity and responsiveness of the network, not a mystical field hovering over it.

This is where the historical movement away from telepathy becomes more than a change of explanation. It becomes a change of *attitude* toward agency. Humans have a deep habit of attributing intention to any coherent motion. A flock is coherent, therefore it must be intended by something. Early naturalists sometimes extended intention upward: the group intends. The modern mathematical approach, at its best, does not deny intention at the level where it is real—animals do intend, in the sense that they have goals and preferences—but it refuses to smuggle intention into levels where it is unnecessary. It asks how far you can go with perception, reaction, and feedback before you invoke anything else.

There is a moral sting in that refusal. "Telepathy" is flattering. It makes nature feel enchanted and, in a way, kinder: the group is unified, harmonious, imbued with a shared spirit. Local-rule explanations can feel colder, like bureaucracy: each bird follows rules, the flock is an outcome, no one is in charge. But the coldness is partly an illusion. A world built from local interactions is not less beautiful; it is often more so, because it gives beauty a mechanism without draining it of wonder. When you watch a flock now, you can hold two images at once: the smooth, sculpted motion at the scale of the whole, and the frantic, limited, bodily labour at the scale of each individual, adjusting to a few neighbours while buffeted by wind and fear.

An anecdote makes the point more concrete. During the Second World War, radar operators and anti-aircraft crews sometimes reported "phantom" objects moving in coordinated ways on early displays, later traced to flocks of birds. The technology was crude, the interpretation uncertain, and the sky full of anxiety. Here was a literal apparatus for detecting distant objects, and it too could be fooled by the signature of collective motion. The machines, like human eyes, saw something coherent and assumed it must be a single entity. Only later did better understanding of both radar and bird behaviour make the confusion legible. The episode is a reminder that telepathy is not merely an old superstition. It is a cognitive default: coherence implies unity.

Mathematical biology did not eliminate that default by scolding it. It eliminated it by giving coherence a new kind of cause. The cause was not an occult channel of communication. It was local sensing married to feedback, played out over time in a network. Once that causal language exists, you can ask sharper questions. How much noise can a swarm tolerate before it falls apart? How many neighbours are needed for cohesion? How quickly must individuals respond to maintain a tight turn? Under what conditions do local interactions produce multiple stable patterns, like milling, streaming, or fission into subgroups? These are questions that do not fit inside a telepathic story because telepathy, by definition, has no constraints. It can explain anything and therefore predicts nothing.

One of the most consequential constraints is that bodies have limits. A bird cannot turn instantly; it has inertia, muscles, reaction times. Its eyes have a field of view; its brain can track only so many moving points. Those limits are not embarrassments to be ignored. They are the joints where explanation grips the world. A flock's grace is partly the grace of a system that has discovered, through evolution, how to operate near the edge of its own stability: responsive enough to evade predators, stable enough not to dissolve into collisions.

Telepathy also faded because scientists became less willing to treat "the group" as a special kind of organism with its own mind. There are cases where the group really does behave like an organism—an ant colony regulating temperature, a slime mould forming a migrating slug—but even there the mind-like properties emerge from interaction, not from a ghost in the collective machine. The phrase *emer-*

gence can sound like a slogan, but its everyday meaning is simple: a pattern appears at a large scale that is not explicitly programmed at the small scale, yet is fully produced by it. A traffic jam emerges from drivers pressing pedals; a market price emerges from buyers and sellers; a rumour emerges from people repeating gossip with small modifications. In each case, the emergent pattern can feel like an actor. It "moves", it "wants", it "forces" behaviour. The scientific task is to enjoy that actor-like quality without being hypnotized by it.

Perhaps the most telling sign that telepathy is gone is how modern researchers talk about these systems. They speak in the language of fields, correlations, interaction networks, phase transitions, information transfer. They build models that can be wrong in precise ways, and then they go looking for the kinds of data that would prove them wrong. The romance has not vanished; it has moved. The romance now lies in the audacity of claiming that a sky full of starlings might be understood with the same intellectual tools used for magnets, fluids, and crowds.

There is still, in the best moments, a place for the old astonishment. Stand again at the edge of the roost. The flock darkens, tightens, peels. A single bird does not know the flock. It knows only a small neighbourhood, a handful of bodies whose movements it must track and anticipate. Yet the flock holds together, not by thought-transference, but by an unceasing cascade of tiny perceptions and adjustments, each one ordinary, each one insufficient on its own, and together enough to carve a living shape out of open air.

7.2 Eyes on the Many

A flock in flight is an optical trick. From a distance it looks like a single thing, almost smooth: a sheet that folds, a ribbon that twists, a punctuation mark against the sunset. Up close it is the opposite of smooth. Birds cross in front of birds, wings flash and vanish, bodies overlap, and the whole mass keeps moving while the camera shakes in the wind. The scientist's problem is not to admire the pattern but to *count* it, frame by frame, without losing anyone.

That sounds like a narrow technical challenge, the sort of fussy work that belongs in an appendix. In practice it is the hinge on which the modern science of collective behaviour turns. Once you can recover the trajectories of hundreds or thousands of individuals—their positions and velocities through time—questions that used to be philosophical become measurable. How many neighbours does an animal respond to? How quickly does a turn propagate across a group? Is information passed like a ripple or smeared like smoke? Does the group have a dense core and a loose edge, or is it more like foam, rearranging but staying statistically uniform? Each of these depends on being able to say, at time t, where each animal is, and at time $t + \Delta t$, which one it became.

Until surprisingly recently, that was out of reach in the field. A human with binoculars can follow one bird. A team with stopwatches can time a few events. But a flock's essential feature is that it is many. The moment you try to watch all of them, the animals hide one another, and even if you could see every dot, you could not keep their identities straight as they weave and swap places. In the language of computer vision, the obstacles have blunt names: *occlusion* (one object blocking another), *correspondence* (matching an object in one frame to the same object in the next), and *calibration* (knowing what your pixels mean in three-dimensional space). Each of those words conceals a headache.

It is tempting to imagine that technology solved the problem by brute force: faster cameras, bigger memory cards, more computing. That helped, but the deeper change was conceptual. Researchers began to treat the swarm not as an untrackable blur but as a system whose structure could be reconstructed indirectly, by combining multiple imperfect views into a single coherent geometry. The approach resembles how a brain reconstructs depth from two eyes. It also resembles the work of early cartographers: no one stands above the landscape, so you stitch together partial perspectives until the terrain emerges.

A concrete example makes the method vivid. In the late 2000s, a group of Italian researchers associated with the StarFlag project filmed starling flocks near roosting sites using synchronized cameras placed at known positions. The scene looks almost comic in photographs: tripods and lenses aimed at an ordinary patch of sky, as if trying to catch ghosts. But the sky is exactly where the geometry is easiest. Against a bright background, each bird becomes a dark point. Two cameras, separated by a baseline, see that point at slightly different pixel coordinates. From those two lines of sight, and careful calibration, you can triangulate the bird's three-dimensional position. Repeat for every bird you can detect, and you get, for a single moment, a three-dimensional "point cloud": the flock as a constellation of dots.

That is already an emotional moment for a scientist, because it turns a living spectacle into something you can plot. But point clouds are only snapshots. The real prize is motion: the sequence of point clouds linked into individual trajectories. That linking is where the trouble intensifies.

Imagine holding a deck of cards and flicking them into the air. In one frame you might see 200 cards. In the next frame you see 200 cards again, but many overlap, many swap relative order, and any given card has moved a little. If you do not label the cards beforehand, you must infer identity from continuity: objects tend not to teleport, so the best match is the one that moved a plausible distance in a plausible direction. That logic works well when objects are sparse. It fails when the density is high and motions are correlated, which is exactly the regime of a flock. Two birds can be equally plausible matches for a dot in the next frame; a brief occlusion can cause you to lose track; and once you make one wrong match, the error can cascade forward in time like a misheard name passed along a line.

Older tracking approaches often tried to be "greedy": match each object to the nearest candidate in the next frame, then move on. In dense swarms, greed is a vice. The better strategy is to solve a global puzzle. Instead of asking, for each bird, "Where did you go next?", you ask, "What assignment of identities across the entire flock produces the most consistent set of trajectories?" Consistency can include smoothness (no wild zigzags), speed limits (birds cannot exceed certain accelerations), and mutual avoidance (two birds cannot occupy the same point). This turns tracking into an optimization problem, closer to scheduling flights or fitting a model than to manually following dots.

Alessandro Attanasi and colleagues introduced one influential step in this direction with a three-dimensional tracking method called GReTA, published in 2013, which tackled the identity problem by combining recursive strategies with global optimization. The key is that the algorithm does not pretend every frame is equally easy. It looks for stretches of time where the data are clean enough to establish reliable correspondences, then uses those as anchors to bridge more ambiguous intervals. It also breaks the big problem into smaller ones, a divide-and-conquer approach that keeps computation feasible when the number of individuals climbs. The details can feel technical, but the intuition is friendly: if you cannot track everyone perfectly all the time, track many individuals very confidently some of the time, and use the structure of motion to fill in the rest.

There is a historical echo here. Before high-speed imaging, astronomers faced a similar identity problem when charting crowded star fields. When points of light are dense and your instrument imperfect, you mislabel stars, invent stars, or lose them. The solution was not merely a better telescope; it was methods for cross-validation, redundancy, and calibration. Swarm tracking borrowed that ethos. A single camera can lie to you in systematic ways. Two cameras can contradict each other in useful ways. Three cameras, carefully synchronized, can turn contradiction into constraint.

Calibration is the quiet hero in this story. To reconstruct 3D positions, you need to know the cameras' intrinsic parameters (how the lens maps rays to pixels) and extrinsic parameters (where the camera sits in space and where it points). Small errors become big at distance. In starling studies, the flock might be tens or hundreds of metres away, so a tiny angular misestimate can shift reconstructed positions by metres. Researchers therefore use calibration targets and careful procedures, often in the same field conditions as the flock itself, and they treat uncertainty as part of the analysis rather than an embarrassment to be hidden.

Once you have trajectories, the swarm stops being a blur and becomes a network of interactions. That word *network* is easy to abuse, so it helps to define it concretely. A network is a set of nodes (individual animals) and edges (relationships). The relationship could be "is within two metres," or "is among the nearest seven neighbours," or "is in the visual field," or "shows correlated turning." The network can change in time as individuals swap neighbours. The central insight is that you can infer something about the rules of behaviour from the geometry and dynamics of these networks.

For example, if interaction is metric (fixed radius), then when density decreases, the number of neighbours within the radius should drop. If interaction is topological (fixed number of neighbours), then the metric distances to those neighbours will expand and contract with density, but the neighbour count stays roughly stable. With trajectories and reconstructed positions, you can test which picture better fits the data. This is not mind-reading. It is rule-reading, and the rules are written in distances, angles, and time delays.

A more delicate inference concerns anisotropy, the fact that animals do not perceive equally in all directions. Birds have eyes on the sides of their heads; many fish have blind spots; insects may have compound eyes with uneven resolution. If you map where an individual's nearest neighbours tend to be located, relative to its direction of motion, you often see an exclusion zone: few neighbours sit exactly in front or exactly behind, because those positions are either hard to see or dangerous for collision. Such anisotropies, measured across many individuals, become signatures of perception and avoidance. They help distinguish *alignment because of aerodynamics* from *alignment because of visual following*, or at least constrain the space of plausible mechanisms.

Correlation analysis provides another window. When the positions and velocities of individuals are known, you can compute how deviations in one part of the flock relate to deviations elsewhere. In a starling flock, Cavagna and collaborators (2010) reported correlations that scale with flock size, suggesting that the influence of local perturbations can extend across the whole group. This does not mean each bird watches the whole flock. It means that the network is connected and responsive in such a way that the effects of local interactions do not die out quickly. The group can behave as if it has long-range awareness because its short-range interactions are arranged, through motion and density, into a system capable of transmitting influence efficiently.

A small but telling detail is the timescale. Human intuition about "instantaneous" is coarse; anything faster than a few tenths of a second can feel like simultaneity. Birds, however, operate at higher temporal resolution. Their reaction times can be tens of milliseconds, and the propagation of a turn through a flock can unfold on a timescale that defeats casual observation. High-speed cameras, recording at dozens or hundreds of frames per second, slice time finely enough to reveal order: not perfect simultaneity, but a sequence of responses, a wave with a speed that can be measured and compared to models. The romantic mystery of "the flock turned as one" dissolves into a more interesting statement: the flock turned as a coordinated medium with a measurable propagation velocity.

The technology also changes what counts as a "natural" observation. Earlier naturalists tended to focus on the most dramatic events: predator attacks, massive roosting displays, the spectacular and rare. Modern tracking allows attention to shift to the ordinary: the quiet maintenance of spacing, the subtle adjustments that prevent collisions, the slow drift of density from core to edge. It is in these mundane details that rules are most visible, because they recur thousands of times per minute. A dramatic attack is like a stress test; ordinary flight is like the baseline physiology. Both matter, but the baseline is where you can most reliably infer the system's default settings.

The same datasets that support biological inference also force philosophical humility. Even with sophisticated tracking, you do not observe "rules" directly. You observe trajectories and infer rules that are compatible with them. Multiple sets of rules can produce similar group patterns, just as multiple recipes can yield similar breads. Technology narrows the space of plausible recipes, but it rarely yields a single answer without assumptions. That is not a weakness; it is honest science. It invites careful experiments: change lighting, change density, compare different species, or exploit natural perturbations like wind gusts and predator passes.

There is also an ethics of attention embedded in the technology. Tracking many individuals at once can tempt researchers to treat individuals as mere particles, ignoring welfare, stress, and context. The best field studies resist that temptation by being minimally invasive. Multi-camera reconstruction of birds in open air is, in this sense, a gentle method: it watches without touching. In other contexts, telemetry tags and implanted sensors provide richer data but impose costs on animals, and those costs must be justified. The swarm becomes legible partly because we have learned how to watch more carefully, not merely more aggressively.

A final, human-scale anecdote captures the mixture of ingenuity and patience that this work demands. Field teams filming starlings near roosts often have to predict where the birds will perform, choose camera placements before the flock arrives, and then wait in cold air with expensive equipment pointed at nothing. When the flock appears, the "experiment" lasts minutes, and everything that matters happens too quickly for human correction. A mistimed start, a slightly drifting synchronization, a lens fogging at the wrong moment, and the night yields only beautiful footage, not usable trajectories. This is the opposite of armchair modelling. It is a kind of disciplined gambling with nature.

When it succeeds, the reward is not just a set of pretty plots. It is the ability to replace vague metaphors with quantified relations: neighbour ranks, interaction ranges, correlation lengths, response delays. The swarm, once treated as a single mysterious actor, becomes a society of bodies whose mutual constraints can be written down. In the raw data, individuality is still there—slight differences in speed, tiny hesitations, local density pockets—but it is individuality seen through a new lens: not as noise to be averaged away, and not as personality to be romanticized, but as the microscopic texture from which the macroscopic pattern is continuously woven.

7.3 The Virtual Aquarium

There is a particular kind of confidence that comes from measurement. Once you can reconstruct a flock as thousands of points in space and watch those points evolve in time, the world feels less enchanted and more legible. Yet measurement alone leaves a stubborn gap. A dataset can tell you *what* happened—how quickly a turn propagated, how tightly neighbours clustered, how the density pulsed—without telling you *why* it happened. In collective behaviour, "why" is rarely a single cause. It is a recipe: a handful of small tendencies at the level of individuals, coupled through the network of who sees whom. The only honest way to test a recipe is to cook it.

That is what agent-based modelling does. An *agent* is simply an individual in a simulation: a digital bird, fish, ant, or person, endowed with a small set of rules and limited information. "Agent-based" means the model is built from the bottom up. Instead of writing an equation for the whole flock, you write instructions for each agent: what it senses locally, how it updates its direction, how it avoids collisions, how much randomness (noise) it suffers, and how strongly it follows any preference. Then you let thousands of these agents run in parallel and see what appears at the group level.

The word "virtual" is not decorative. Done well, this is not animation. It is a laboratory that can be rewound, perturbed, and repeated without harming any animal. You can change one assumption at a time and watch the consequences, the way a chemist changes one reagent. What counts as success is not that the simulation looks vaguely like nature. Success is when the same local rules reproduce multiple quantitative features of real groups: their shapes, their fluctuations, their response times, their propensity to split or remain cohesive under stress. A model that only gets the silhouette right is often the least trustworthy one, because silhouettes are easy to fake.

The historical moment when this approach entered popular consciousness is unusually clear. In 1987, Craig Reynolds, then working in computer graphics, described a simple distributed model for flocking that he called *boids* (short for "bird-oid objects"). The motivation was not initially biology; it was animation. Filmmakers wanted crowds and flocks that moved believably without hand-animating each individual. Reynolds' move was to refuse the temptation of choreography. Instead of scripting global motion, he gave each boid a small perceptual neighbourhood and a few behavioural tendencies. The flock emerged as a side effect.

The boids idea has been retold so many times that it can sound like a folk tale: three rules and suddenly you get birds. The real achievement is subtler. Reynolds showed that if you choose rules that are local, continuous, and symmetric (no bird is privileged), then you can get stable, lifelike collective motion without any central coordination. That was not just a trick for cinema. It was a demonstration that the telepathy illusion can be replaced by computation: not the computation of a master planner, but the computation distributed across many small decision-makers.

The canonical boids rules are usually named *separation*, *alignment*, and *cohesion*. These names can mislead, because they sound like moral virtues rather than vector calculations. It helps to translate them into physical intuition.

Separation is personal space: if someone is too close, move away. Alignment is conformity: match your heading with neighbours so you are not swimming upstream while everyone else moves downstream. Cohesion is social gravity: do not drift too far from the group; steer gently toward where your neighbours are. None of these requires global knowledge. Each can be computed from what lies within a limited sensory radius.

Placed in code, the rules become forces (or, more precisely, steering vectors) that are combined into a single acceleration. The rules are not mutually compatible in every moment. Separation pushes away, cohesion pulls in. The flock's smoothness comes from how these competing tendencies balance, frame after frame, so that no one ever quite collides and no one ever quite escapes. The beauty is not in the individual rules; it is in their compromise.

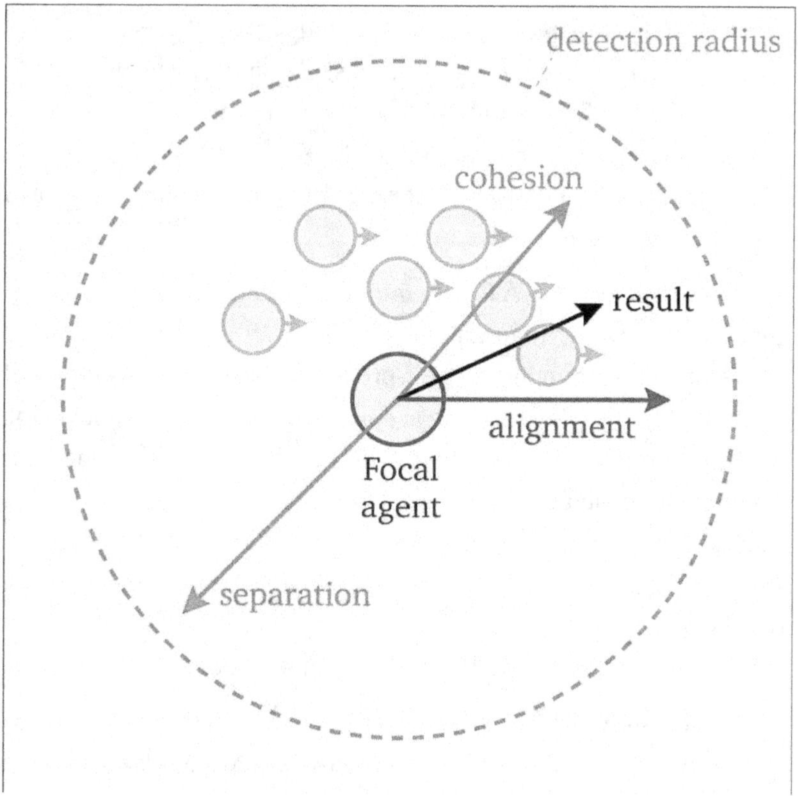

The diagram is deliberately simple. Real implementations vary: the detection radius may be replaced by a fixed number of nearest neighbours; the rules may be weighted; speed may be capped; turning may be limited by inertia; noise may be added. But the picture captures the key intellectual move. Collective motion is written as the vector sum of local influences. Nothing in the code says "form a flock." Flocking is what happens when personal space, conformity, and group attachment are tuned so that they can all be satisfied most of the time.

Reynolds' boids also helped to clarify a broader methodological point: when studying swarms, *realism* is not a single axis. A model can be biologically unrealistic in its details and still be scientifically valuable if it captures the right causal skeleton. Conversely, a model can be stuffed with anatomical detail and still miss the essence if it fails to reproduce the correct group-level phenomena. The virtual aquarium is useful precisely because it lets you separate "must-have" ingredients from decorative ones.

A few years after boids, another minimal model became a workhorse for scientists rather than animators. In 1995, Vicsek and collaborators published a model of self-propelled particles that align with their neighbours while moving at constant speed, with a controllable amount of random perturbation. The simplicity is almost shocking. There is no explicit cohesion term pulling agents together, no elegant collision-avoidance geometry, no model of vision. And yet, as you vary density or noise, the system undergoes an abrupt change from disorder to collective motion. The group *chooses* a direction, even though no particle is told which direction to choose.

This matters because it gives swarm science something like a phase transition, a concept borrowed from physics. Water freezes not because each molecule decides to be ice, but because local interactions make some states stable and others unstable as temperature changes. In the Vicsek model, "temperature" is played by noise. High noise destroys alignment; low noise allows alignment to persist and spread. The model is not a metaphor. It is a concrete set of update rules that can be studied mathematically, simulated quickly, and compared to data.

The phase-transition idea does something else: it legitimizes the thought that a flock can have *collective states* with their own dynamics. A flock can be in a milling state (circling), a polarized state (moving coherently), a disordered state (swirling without net motion), and it can switch between them as conditions change. Those switches do not require new intentions. They can be the natural consequences of parameters drifting: density, noise, speed, interaction range. For an animal group, the parameters are influenced by biology and environment: predator proximity, visibility, hunger, fatigue. In a virtual aquarium, they are sliders. A scientist can move them and watch the system respond.

The temptation, once you have such a model, is to stop there: to declare that flocking has been explained. But an honest model does not end explanation; it raises sharper questions. Which parameters correspond to real sensory limits? Do individuals align with all neighbours in a radius, or with a fixed number of nearest neighbours? Does the interaction depend on metric distance, topological rank, or something closer to what the animal actually sees, such as angular size on the retina? Does information propagate diffusively, like random walks, or in wave-like modes? Models become meaningful when they are placed under the discipline of such questions.

Here is where the virtual aquarium closes the loop with measurement. Consider the earlier finding that starlings appear to interact with a fixed number of neighbours. In code, that is a small change: replace "all neighbours within R" with "the closest k neighbours," perhaps $k \approx 6$ or 7. The consequences, however, are not cosmetic. A topological neighbourhood can stabilize cohesion under changing density. It can change the statistics of who influences whom. It can alter how perturbations propagate across the interaction network. In other words, a single modelling choice creates testable predictions. If the predictions match real flocks across multiple conditions, confidence in the underlying rule grows.

Agent-based models are also a way to explore something that field data often conceal: heterogeneity. Real groups are rarely composed of identical agents. Some individuals are hungry, some satiated; some are experienced, some juveniles; some have information, some do not. A particularly elegant set of demonstrations came from work by Iain Couzin and Jens Krause around 2005, showing that a small fraction of "informed" individuals can guide a moving group without any explicit signals. In the model, informed agents have a preferred direction (toward a goal), while uninformed agents merely align and cohere. If the informed minority is too small, it is swallowed by the crowd and the group wanders. Above a threshold, the preference of the informed agents biases local interactions just enough to steer the whole group.

The dinner-party version of this result is disarming: leadership can be invisible. There may be no gestures, no calls, no dramatic assertions of authority. The group can move as if it has chosen a destination even when most members have no idea where they are going, simply because local alignment allows a small directional bias to spread. The scientific version is more precise: the model identifies conditions under which consensus emerges, conditions under which the group splits, and how these outcomes scale with group size and with the fraction of informed individuals. Those are questions that are almost impossible to answer by intuition alone, because the dynamics are nonlinear. The virtual aquarium gives intuition something to lean on.

At this point, it is natural to ask whether modelling a swarm as thousands of interacting agents is the only route to understanding. It is not. There is another style of theory that aims for universality: instead of describing each agent, it describes the flock as a continuous medium, like a fluid with its own rules of motion. John Toner and Yuhai Tu's work in 1995 provided a foundational framework of this sort. Their equations do not care whether the underlying agents align by averaging headings or by following nearest neighbours. They focus on what large-scale properties must follow from a few symmetries and constraints: self-propulsion, local alignment, and the fact that the system is out of equilibrium (it consumes energy to move). The result is a kind of hydrodynamics for flocks, predicting collective modes and long-range order even in two dimensions, where equilibrium analogues would forbid it.

It might seem that such continuum theory makes agent-based modelling unnecessary. In practice the two approaches are companions. Continuum theories tell you what kinds of large-scale phenomena are generic and what kinds are fragile. Agent-based models tell you which microscopic rules can plausibly generate the effective parameters that appear in the continuum equations. If the continuum theory predicts a wave-like mode of turning, the agent-based model can reveal what micro-level ingredients are needed to sustain it: perhaps an inertia-like tendency to maintain direction, or a delayed response, or a particular form of coupling. Conversely, if a certain agent-based model generates patterns not observed in nature, continuum theory can help explain why: an instability that real animals suppress, a symmetry that is broken in real perception, an unrealistic noise model.

The phrase "virtual aquarium" captures another, more intimate strength of simulation. Aquariums are places where you can watch a fish from multiple angles, under controlled light, with the background stripped away. A virtual aquarium does something similar, but for causality. You can remove predators entirely and ask what structure remains. You can add a single informed agent and see whether it is amplified or ignored. You can introduce a temporary sensory blind spot and see whether the group becomes more collision-prone. You can test interventions that would be unethical or impossible in the real world.

The greatest risk is mistaking plausibility for truth. A simulation can feel persuasive because it is dynamic and visual. It moves; it looks alive; it produces the very patterns that inspired the question. But a convincing animation is not the same as a confirmed mechanism. The most responsible modelling culture treats simulations as arguments that must be checked, not as demonstrations that end debate. This is why comparison to data is so central. It is also why modern modelling increasingly looks beyond the obvious: not just whether the group forms a flock, but whether the distribution of nearest-neighbour distances matches, whether turning waves attenuate as observed, whether correlations scale with group size in the same way, whether the group's shape statistics resemble those measured in the field.

When modelling succeeds in this stricter sense, it changes the emotional tone of explanation. The spectacle of a flock remains, but the spectacle is no longer taken as evidence of mystery. It becomes evidence of compressibility: the idea that a complicated moving world can be generated by a surprisingly compact set of rules. That compressibility is not a denial of life's richness. It is a clue about where richness resides. In a swarm, much of the apparent sophistication is not stored inside any one brain. It is stored in interaction.

The most satisfying moment in a virtual aquarium is often not when the flock first coheres, but when it fails in an informative way. Increase noise slightly and the group fragments. Reduce cohesion and the swarm stretches into filaments that snap. Increase separation too much and the agents repel into a gas. Add a few stubborn informed individuals with conflicting goals and the group splits, not because it cannot decide, but because local rules make compromise structurally unstable. These failures are gifts. They mark the boundary between collective intelligence and collective confusion, and boundaries are where mechanisms show themselves most clearly.

A flock that can turn without collision, hold together under attack, and still remain flexible enough to respond quickly is a delicate achievement. In code, that delicacy appears as a narrow region of parameter space where everything works. In nature, it appears as evolution's patient tuning: sensory systems, reaction times, body mechanics, and simple behavioural heuristics shaped until a moving group can be both stable and responsive. The virtual aquarium does not replace the living world; it gives it a second stage, where hypotheses can be made to perform under bright lights, with every hidden string pulled into view.

Chapter 8

Silicon Mimics and Mortal Limits

We have learned to forge silicon workers that act as one, translating the dance of bees into search-and-rescue algorithms that require no central commander. Yet, the same simple rules that drive success can lock biological and mechanical swarms alike into fatal feedback loops. Ultimately, these mimics teach us that collective intelligence is not a lesser version of human thought, but a distinct, alien form of mind requiring its own ethical framework.

8.1 The Algorithmic Hive

A lone robot in a collapsed building faces a familiar dilemma: it cannot see through concrete, it cannot climb every void, its radio fades behind rebar and wet cement, and its battery does not care about heroism. Add a second robot and new troubles appear immediately. They collide, jam in the same doorway, repeat each other's work, and drown the radio channel in chatter. Add a hundred and the problems become less like engineering and more like ecology: congestion, interference, rumor, panic, waste.

The hard-won lesson of swarm biology is not that groups are magically wise, but that *coordination is possible when each individual is stupid in the right way.* "Stupid" here means something precise: each agent reacts to what it can sense locally, follows rules that fit in a small controller, and never assumes it possesses a global map. That style of intelligence is easy to admire in ants and bees because it is alive and ancient. Engineers admire it because it is robust. A centralized system fails like a chandelier: one snapped chain, and the whole thing drops. A decentralized system fails like a bush: you can break branches and still have leaves.

Robotics inherited its first instincts from the chandelier. Early mobile robots were often designed as solitary explorers with a single brain: a planner, a map, a path, a sequence. When multiple robots were used, they were frequently treated as remote limbs of a single computation, coordinated by a base station or a command vehicle that watched everything and assigned tasks. That works in a clean warehouse with Wi-Fi and fiducial markers on the floor. It works much less well in smoke, dust, and noise, where the "commander" cannot hear half its team and the team cannot hear the commander.

The biological translation begins with a shift in what counts as a plan. Instead of a detailed script, you give each robot a set of imperatives: don't collide; don't drift too far from your neighbors (unless you must); align your motion when alignment is helpful; keep moving; report what you find when you can. Imperatives are not goals in the human sense. They are constraints, like the rules of grammar. Out of constraints, surprisingly often, come coherent sentences.

The most influential cartoon of this idea in modern computing did not start with insects at all. It started with birds that were not birds. In 1987 the computer graphics researcher Craig Reynolds wrote a program to simulate flocking for animation. Each "boid" (bird-oid) followed three simple rules based on nearby neighbors: separation (avoid crowding), alignment (match heading), and cohesion (steer toward the group). Nothing in the code said "form a flock." Yet flocks formed: they split around obstacles and rejoined, stretched into ribbons, tightened into knots, and swung as if tugged by a single invisible hand.

For robotics, Reynolds's insight is less about the beauty of collective motion than about a general method: start with local interactions that are easy to compute, then tune them until the global pattern emerges. The elegance is deceptive. Each rule is local, but the effect is not merely local. If a robot steers away from collisions, it is not only protecting itself; it is shaping traffic. If it tries to remain near peers, it is not only avoiding loneliness; it is maintaining a network through which information can flow. If it aligns, it is not only being polite; it is dampening chaos.

Engineers often describe these rules in terms that sound like physics: virtual forces, potentials, springs. Imagine every robot pushing away from nearby robots at short range (repulsion) and pulling toward them at longer range (attraction). The math is convenient. You can plot the forces, prove the system will not crash into itself under certain conditions, and adjust parameters like an audio engineer adjusts equalizers. But beneath the math sits a biological commitment: you are trying to get many limited creatures to share a space without a traffic cop.

Search-and-rescue sharpens the requirement. After an earthquake, a building becomes a three-dimensional maze whose corridors reconfigure as dust settles. GPS is useless indoors; cameras fail in darkness or smoke; maps are wrong because walls have moved. A centralized "overseer" would need constant, reliable updates from each robot to coordinate coverage. That update stream is exactly what the environment refuses to allow.

So swarm robotics embraces a different bargain. Each robot becomes responsible for a small patch of the world, in the only sense it can afford: it reacts to nearby obstacles, maintains a loose relationship with neighbors, and collects data that can be shared opportunistically. If the radio link drops, the robot still functions. If half the robots are crushed or drained, the remainder still explore. If a robot is trapped, it is an individual tragedy but not necessarily a system failure.

This is where the word *decentralized* can mislead. Decentralization does not mean "no structure." It means structure without a single point of command. In a human emergency response, teams form, split, and re-form; leadership shifts as expertise becomes relevant. A firefighter with thermal vision becomes the center of a small decision loop for a moment. A paramedic becomes the center for a different moment. Swarm robotics tries to achieve a crude version of the same fluidity using local rules rather than social negotiation.

The details look mundane at first: timers, thresholds, and short-range messages. A robot in a hallway might broadcast a simple "frontier" signal: *I have found an unexplored branch.* If nearby robots receive multiple such signals, they choose one probabilistically, or based on distance, or based on how long it has been waiting. That choice mechanism matters enormously. Too greedy, and they all chase the nearest glittering unknown, leaving other areas untouched. Too cautious, and they wander without committing, like indecisive shoppers in a supermarket aisle.

A popular tactic is to mimic the randomness found in biological foraging. Animals do not always move in smooth, planned lines; they jitter, hesitate, and occasionally make apparently irrational long moves. In algorithmic form, those occasional long moves can prevent a group from getting stuck. The jargon term is a "Levy flight," a movement pattern in which many steps are short but some are very long. The underlying intuition is easy: if you only take short steps, you keep revisiting the same neighborhood. A few long steps are like opening windows in a sealed room.

But random motion alone is not a rescue strategy. It must be braided with repulsion and attraction: spread out enough to cover ground, stay connected enough to share discoveries, avoid clumping, avoid disconnection. Engineers speak here of *coverage* and *connectivity* as if they were separate checkboxes. In practice they fight each other. Maximum coverage pushes the swarm to disperse until radios fail; maximum connectivity pulls them together until they waste time overlapping. Biological swarms live inside this tension continuously. Ants spread into branches of terrain, yet they keep trails that allow resources to flow back. Birds spread to forage, yet they keep enough cohesion to reduce predation risk. The "right" point is not universal; it depends on the environment's hostility and the task's urgency.

One reason engineers love swarms is that they allow a kind of graceful ignorance. A single robot asked to map a building must solve a grand problem: simultaneous localization and mapping, or SLAM, in which it estimates its position while building a map. SLAM is powerful, but in messy environments it is brittle. A swarm can do something less heroic. Each robot can keep a local, approximate map-more like a set of memories than a blueprint-and the group can still achieve useful outcomes: finding victims, identifying passable corridors, or detecting hazardous gas. The collective behavior substitutes redundancy for certainty.

The historical anecdote here belongs not to a battlefield or disaster zone but to a lab bench in Massachusetts, where a group of researchers made the idea almost comically concrete. At Harvard, Radhika Nagpal's group built large swarms of tiny robots called Kilobots-simple, cheap devices designed to show how collective behavior could be programmed at scale. Each unit had minimal sensing and limited communication. They could not do much alone. Together, they could form shapes, coordinate movement, and demonstrate that simple local algorithms could produce global patterns reliably even when individual robots were noisy or miscalibrated.

The charm of Kilobots is that they strip away romance. These are not sleek humanoids. They are proof that coordination does not require sophistication at the individual level. If you can make a thousand small machines agree on a shape with only local communication, you can begin to imagine a thousand small machines agreeing on something more consequential: how to sweep a field for chemical leaks, or how to form a distributed radio relay inside a tunnel.

Yet rescue robotics raises a brutal question: can the principles that make a lab demo work survive gravity, rubble, and the indifferent complexity of real terrain? To answer that, engineers needed not only clever algorithms but a different view of communication and control. In a real disaster site, the environment itself can become part of the coordination system, just as it is for termites.

Termites do not carry blueprints. A termite mound is not built because one termite knows the plan. It is built because each termite responds to local cues: the presence of a pellet here, humidity there, a chemical marker somewhere else. The environment stores information in the form of partial structures, like a note left on a table. Biologists call this *stigmergy*: coordination through the traces that work leaves behind.

In 2014, Justin Werfel, Kirstin Petersen, and Radhika Nagpal presented a termite-inspired robot construction team that made stigmergy tangible. The robots were not chatting about architecture. They moved blocks and responded to what they encountered: a block here implies another block should be placed there; a partial wall creates a cue for the next layer. The "plan" lived in the rules and in the evolving structure itself. Remove a robot and the construction continued. Replace a robot with a slightly different one and it could still participate if it followed the same local conventions. It was not a perfect imitation of termites—nothing in steel and plastic truly is—but it showed a direction: use the world as memory.

Rescue scenarios are full of opportunities for environmental memory. A robot can drop a small marker—physical or digital—that says *this corridor is blocked* or *this room has been searched.* In simulation it might be a virtual pheromone field. In reality it might be a QR-like tag, a reflective patch visible to infrared, a small radio beacon, or even a chalk mark if humans are part of the team. The point is not the gadget; it is the principle: don't force every robot to remember everything. Let the environment help.

This shifts the role of communication. Instead of streaming continuous updates to a commander, robots exchange short, local messages and interpret the shared world. The communication becomes more like gossip than like a corporate memo. A robot that finds a victim might broadcast a high-priority signal. Nearby robots might converge to form a relay chain to the outside, because the simplest way to increase radio range in a rubble-filled building is often to place nodes like stepping stones. That relay chain is itself a swarm behavior: robots acting as infrastructure for other robots.

Designers talk about these patterns as if they were obvious, but each has a subtle failure mode. If too many robots rush toward the same signal, they create congestion in narrow passages. If the group forms a relay chain too eagerly, it sacrifices exploration for communication. If it explores too eagerly, it loses the ability to transmit discoveries. Once again the biological analogy helps, not because ants solve everything, but because ants reveal the tradeoffs. A trail is useful, but a trail is also a constraint: once laid, it attracts more traffic, which deepens it, which attracts even more traffic. The same feedback loop that produces efficient paths can also produce stubbornness.

Engineers therefore borrow not only positive rules (follow, align, reinforce) but also negative ones: decay, inhibition, retreat. In algorithms inspired by pheromones, "pheromone" evaporates: old information fades so the swarm can adapt. In robot swarms, a "frontier" signal may time out. A relay chain may be allowed to persist only if it is actively carrying traffic; otherwise the robots dissolve back into explorers. These negative feedbacks are less exciting than cohesion and alignment, but they are often the difference between a swarm that adapts and a swarm that ossifies.

The phrase "algorithmic hive" captures a second translation from biology into code: optimization. Honeybees searching for a new nest site do not merely wander; they compare options, recruit others, and eventually commit. Engineers have built families of algorithms inspired by this style of collective search. In the 1990s, James Kennedy and Russell Eberhart introduced particle swarm optimization, in which candidate solutions "fly" through a mathematical landscape influenced by their own best positions and their neighbors' best positions. Ant colony optimization, developed around the same period by other researchers, uses virtual pheromone trails to bias search toward promising routes.

These are not robotics algorithms in the narrow sense; they are ways of solving hard problems by imitation of collective behavior. But the conceptual bridge matters. When you instruct a swarm of robots to explore, you are also asking them to solve an optimization problem in real time: maximize discovered area, minimize time, minimize energy use, maintain connectivity, avoid risk. Those goals conflict, and no single robot can compute the best compromise. Swarm-inspired optimization offers a way to let the compromise *emerge* through local influence.

It is tempting to imagine that this makes design easy: write down a few rules, and nature does the rest. Real swarms are not so forgiving, and neither are artificial ones. The engineering challenge is the "inverse problem" of collective behavior: you want a global outcome (efficient exploration and reliable reporting), and you must find local rules that produce it under uncertainty. In biology, evolution spends millions of years tuning those rules. In engineering, you do it with simulation, proof, and a kind of humble trial-and-error.

Simulation is indispensable, and also dangerous. A swarm that works in a clean simulator can fail in the real world for reasons that feel petty: wheel slip, uneven floors, dusty sensors, small delays in communication that accumulate into oscillations. Reality injects latency into the nervous system of the swarm. One robot turns a fraction of a second late, another overcorrects, and suddenly a corridor becomes a clog of spinning bodies. Engineers often respond with additional damping: slower reactions, more conservative spacing, simpler decisions. Paradoxically, making each robot *less* eager can make the swarm *more* effective.

When swarm robotics is discussed at conferences, the word *scalability* appears like a charm against evil. A system scales if it works not only for ten robots but for a hundred or a thousand. Biology scales because the rules are local: an ant interacts with a handful of neighbors, not with the whole colony. Swarm engineers mimic this explicitly. They limit message range, cap the number of neighbors each robot considers, and avoid anything that requires a shared global clock. The goal is to prevent the swarm from becoming a committee meeting.

Fault tolerance is the second charm. A rescue swarm cannot assume every unit survives. Some will be crushed. Some will lose sensors. Some will run out of power in the wrong place. Centralized systems treat failures as exceptions; decentralized swarms treat them as normal weather. The price is redundancy. You send more robots than you need, knowing some will vanish. That sounds wasteful until you compare it with the cost of a single exquisite robot that fails.

The most human part of this engineering is deciding what "success" means. In a disaster, perfect coverage is not required; *timely* coverage is. Finding two victims quickly may matter more than mapping the entire structure. Swarm design therefore often includes urgency-sensitive behaviors. Robots may bias exploration toward areas that "look" like human occupancy-open spaces, heat signatures, sounds-while still maintaining enough randomness to avoid missing someone behind an unlikely barrier. They may switch modes: dispersing early, clustering later to act as relays or to escort a human team. These mode switches are a kind of minimal "cognition" at the individual level, but they are still local: a robot switches because its battery dips, because it detects a high-priority signal, because its neighbor density changes.

A useful way to picture the algorithmic hive is as a set of overlapping fields rather than a set of explicit commands. There is a collision-avoidance field that pushes robots apart at short distances. There is a cohesion field that pulls them together at moderate distances. There is an exploration field that pulls toward unexplored frontiers. There is a communication field that rewards staying within relay range. Each robot sits at the intersection of these invisible pressures and takes the simplest possible step that reduces its discomfort. The miracle is that discomfort, shared across many bodies, can become something like strategy.

And yet the miracle is not mystical. It is the product of careful choice about what information each robot is allowed to use. If you give too much, you recreate the chandelier: a dependency on global state, on perfect communication, on synchronized maps. If you give too little, you get a drift of atoms. Biological swarms live between these extremes because their senses and their environments force them to. Artificial swarms must be designed into that constraint deliberately.

The most promising rescue swarms will likely look less like a disciplined platoon and more like a resilient organism: a distributed sensory surface that can be bruised without losing all feeling, a shifting internal network that routes signals around damage, and a capacity to concentrate effort where it matters without needing a single brain to decree that it matters. When the rubble blocks the radio, the swarm does not wait for permission; it rearranges itself until a path for information opens, because that is what its rules make it do.

In that sense the "hive" is not a metaphor of sweetness and industry. It is a metaphor of survival. A hive endures winter not because each bee understands winter, but because the colony has rules that make warmth and repair more likely than cold and collapse. Engineering cannot borrow the bees' bodies or their evolution, but it can borrow the lesson: if you want intelligence under mortal limits, design for local perception, distributed responsibility, and the expectation of loss. The rest is patient work—tuning thresholds, shaping feedback, and accepting that the smartest thing a machine can do in a disaster may be to remain simple enough to keep moving.

8.2 The Death Spiral

A self-organizing system is at its most impressive when it rescues itself from trouble. A few units fail; the rest reroute. A path is blocked; the group flows around it. A signal drops; the network heals. That resilience can feel almost moral, as if the collective has learned humility: no single point of pride, no single point of collapse.

The same ingredients, mixed slightly differently, can become lethal.

The most haunting example is the ant mill, sometimes called a death spiral. Certain trail-following ants navigate by a rule that is usually sensible: keep moving, and prefer the path that smells like other ants. The smell is pheromone, a chemical trace laid down as they walk. In ordinary conditions the rule is a triumph of distributed intelligence. Pheromone lets a colony lay down efficient routes to food without a cartographer. The stronger trail is often the better trail because more ants have tested it. The colony exploits experience without any ant needing to remember much.

Trouble begins when the trail no longer encodes a good route but merely encodes itself.

Picture a moving column of ants in dense vegetation. Something disrupts the line: a gust, a broken branch, a predator, a sudden gap in the pheromone path, or simply the confusion of an obstacle that forces detours. A few ants lose the original trail and begin to follow one another, because following another ant is usually a decent proxy for following the colony's intent. If the geometry is unlucky, that local following closes into a loop. Once the loop exists, each lap deposits more pheromone on the same curve. The curve smells increasingly like *the right way*, not because it leads anywhere, but because it has become the most intensely marked option. New ants arriving at the edge of the loop encounter a strong trail and join it. The circle tightens and stabilizes. The group marches, and marches, and marches.

William Beebe, the naturalist, described such behavior in 1921 after observing army ants apparently trapped in a rotating ring. The image is difficult to shake because the ants are not panicking in any human sense. They are functioning. Each individual is doing something close to what it was built to do: follow local cues, keep the body moving, maintain contact with the collective flow. The catastrophe is not a breakdown of the rules; it is the rules succeeding too well in the wrong landscape.

That is what makes the ant mill more than a curiosity. It is a diagram of a general failure mode: a positive feedback loop that has lost its brakes. Positive feedback is the kind that amplifies whatever happens to start it. A microphone too close to a speaker produces a scream because a small sound is re-amplified, then re-amplified again, until it saturates the system. In a pheromone trail, the "sound" is chemical. In an economy, it might be rumor. In a social media swarm, it might be outrage. The machinery differs, but the logic is the same: local reinforcement plus delay in correction can create a stable pattern that is globally disastrous.

In an ant mill the global disaster is exhaustion. Ants can die from the effort if the loop persists long enough, especially in harsh conditions. Even when they do not die, time is lost, and time is the one resource a foraging swarm never truly has. A trail-following system is designed to convert motion into meaning: *where others went recently is probably where I should go.* The mill converts motion into meaninglessness: *where others went recently is where they went recently.*

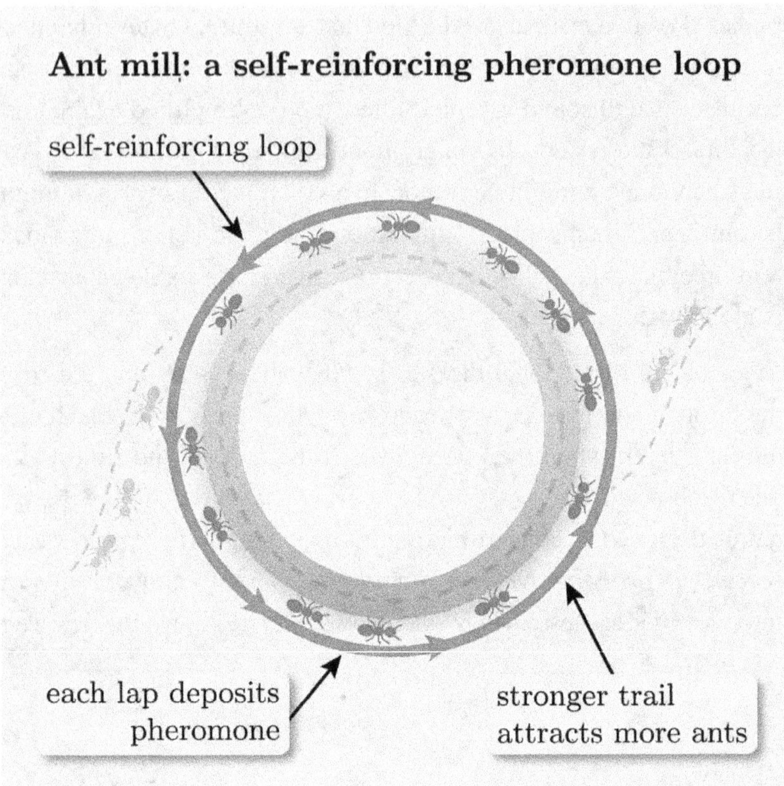

The geometry in the diagram is deceptively clean, a circle drawn with a compass. Real ant mills can be messier: spirals that tighten, loops that wobble, arcs that become circles as the chemistry intensifies. But the clean geometry clarifies the essential point. The loop is not imposed from above. It is an emergent trap built from the same tool that normally enables success.

Engineers have their own versions of the ant mill, though they rarely come with the drama of insects dying in a ring. A swarm of robots told to "follow the strongest signal" can lock into following each other's transmissions rather than the source. A fleet of drones told to "maintain cohesion" can clump in a doorway, each unit trying to avoid separation by moving closer, making the jam worse. A set of delivery robots told to "take the shortest route" can create congestion on the same corridor until the shortest route becomes the slowest, and still keep choosing it because the rule is blind to delay. Even simple collision avoidance can, under certain layouts, produce endless oscillations: two robots meet in a narrow passage, both step aside, both re-enter, both step aside again, as if caught in a polite but fatal dance.

The kinship between these failures is not that the agents are badly programmed. It is that the local rules lack a reliable mechanism for negative feedback, the kind that dampens runaway reinforcement. Negative feedback is what a thermostat provides: when the room warms past the target, the heat turns off. In biological swarms, negative feedback often arrives through evaporation (pheromones fade), through saturation (a trail becomes too crowded to be efficient), through switching thresholds (bees stop recruiting when a quorum is reached), or through physical limits (a jam forces rerouting). But those brakes are not guaranteed. They depend on the environment. In an ant mill, evaporation may be too slow relative to reinforcement. Physical congestion does not dissolve the loop because the loop itself is the congestion. The "signal" is the crowd.

The uncomfortable implication is that self-organization is not automatically self-correcting. People sometimes talk about "wisdom of crowds" as if aggregation washes away error. Crowds can also magnify error. If the cost of being wrong is small for each individual but the cost is concentrated at the collective level, the swarm can drift into a cliff edge with no single agent feeling the slope.

A useful way to say this without mysticism is to focus on the information each agent uses. Trail-following ants use a proxy: pheromone strength stands in for "goodness of route." The proxy works because, most of the time, good routes get used more and therefore smell stronger. The ant mill is what happens when the proxy becomes detached from the thing it was meant to measure. The pheromone is no longer "evidence of progress"; it is evidence of pheromone. The swarm has created a self-referential signal, a local sign that points only to itself.

Human institutions have analogous proxies. Popularity stands in for quality; repetition stands in for truth; authority stands in for competence. These stand-ins are not foolish in themselves. They save time. They let a person with limited knowledge take advantage of social information. But when the proxy detaches from reality, the feedback can become vicious. A false rumor repeated often enough begins to look like consensus. A financial asset bought merely because others are buying becomes a bubble. The pattern is the same: reinforcement that feeds on its own output.

What makes swarm failures distinct is that responsibility is smeared. In a centralized system, when the hub fails, one can point to the hub. In a distributed system, failure is often a property of interactions rather than components. Each unit can be correct in isolation and still participate in a collective mistake. That creates not only technical difficulty but moral discomfort. A rescue swarm that wastes time in a loop has not "decided" to waste time, and yet the consequence is time lost where time may mean life.

One response is to try to restore a little centrality: give the swarm a supervisor that can detect pathological patterns and override local rules. Engineers do this in practice, often quietly. Even systems marketed as "fully decentralized" may include intermittent global checks, a base station that can broadcast an emergency reconfiguration, or a higher-level planner that sets boundaries. In a collapsed building, however, the supervisor is often precisely what cannot be relied upon. The radio link is intermittent, the map is incomplete, and the situation changes faster than centralized cognition can keep up.

So the more interesting response is to engineer negative feedback into the local rules themselves, without requiring global awareness. In pheromone-inspired algorithms, this can be done by enforcing decay aggressively: information must be refreshed or it disappears. In robot swarms, one can impose "impatience": if a robot has followed a gradient for too long without evidence of progress, it deliberately injects randomness, like a person leaving a crowded hallway to try a different staircase. Another tactic is to detect loops locally. A robot can keep a short history of its recent turns and infer that it is cycling. It does not need a map of the building to notice that it has turned left every few seconds for the last minute.

There is a subtle psychological analogy here. Humans are skilled at escaping loops because we can name them. "We've been here before" is a powerful sentence. Swarm agents cannot name in that way, but they can be given crude loop-detectors: counters, timers, thresholds that trigger a mode switch. The cost is that they sometimes abandon a good trail too early, mistaking persistence for pathology. The benefit is that they are less likely to die in a circle.

Another brake is heterogeneity: letting agents differ. A uniform swarm can lock into uniform mistakes. If every ant follows the same rule with the same sensitivity, the loop is sticky. If some ants are more exploratory, less responsive to pheromone, or more willing to break alignment, they can seed escape routes. Biology often uses this trick. Colonies contain individuals with different thresholds for action. Some are more excitable recruiters; others are skeptics. In engineered swarms, diversity can be intentional: a fraction of robots may be programmed as "scouts" that prioritize novelty over cohesion, or as "anchors" that prioritize communication stability, or as "auditors" that test whether the strongest signal is actually helpful. The goal is not hierarchy but a kind of internal friction, the social equivalent of non-slip soles.

The need for such friction becomes more urgent when considering adversarial influence. A self-organizing swarm can be manipulated by manipulating its local cues. If ants can be redirected with pheromone-like chemicals, then robots can be redirected with spoofed signals, fake markers, or compromised agents. An attacker does not need to seize a central controller if there is no central controller to seize. It is enough to bias the local rule updates so that the swarm, faithfully following its own imperatives, drifts toward a trap. The ant mill is a natural accident; a malicious loop would be the same structure with intent added.

This is one reason assurance—the attempt to provide credible arguments that a system will behave safely—is unusually difficult for swarms. A single robot can be tested like a product: does it stop at a wall, does it avoid a human, does it obey a command? A swarm is defined by interactions. The dangerous behavior might occur only at scale, only with certain communication delays, only in certain geometries, only when a few agents fail in a particular pattern. Testing every possibility is impossible. Proving correctness in the usual mathematical sense is rarely feasible. The practical approach is closer to engineering in the presence of weather: build in margins, build in brakes, build in ways to fail softly rather than catastrophically.

The ant mill offers a surprisingly concrete checklist for this kind of humility. Ask where the positive feedback lives. Ask what counts as evidence for an agent. Ask how that evidence can become self-referential. Ask what local mechanism could interrupt reinforcement without waiting for global insight. Ask how the system behaves when the environment is not merely noisy but adversarial or misleading. Above all, ask what the swarm is allowed to do when it is uncertain. Many disasters in autonomous systems come from treating uncertainty as an inconvenience rather than as a state that should trigger conservative behavior.

The biological sting is that ants did not "choose" their vulnerability. Their rule evolved because, across the usual range of environments, it works. The rare failure is paid for by individuals, and evolution tolerates a certain casualty rate if the overall strategy is profitable. Engineers do not have that luxury. When human lives are involved, "rare" is not a comforting word. The moral center of rescue robotics is that failure is not merely a cost; it is someone's air running out.

And yet there is a strange hope in confronting the death spiral directly. The ant mill is a failure made visible. It is not a silent bug hidden in code. It is a geometric confession: a loop drawn on the ground by creatures following their logic to its end. Seeing the loop makes it easier to imagine breaking it, not with omniscience, but with small, local permissions to doubt the crowd.

A swarm that can save lives will never be purely trusting. It will need a way to loosen its grip on its own strongest signals, to treat reinforcement as a hint rather than a command, and to remember—in whatever minimal form a machine can remember—that the world is larger than the path it has worn into itself.

8.3 Mind Without a Face

People cannot resist giving a swarm a personality. Watch honeybees hanging in a living curtain from a branch and someone will call them *calm*. Watch starlings knot and un-knot in the evening sky and someone will call them *playful*. Watch army ants flood a forest floor and someone will call them *relentless*. The temptation is even stronger when the swarm is built from machines. A thousand tiny robots moving with uncanny coordination feels like a single creature that happens to be made of parts.

Anthropomorphism is not merely a poetic habit; it is a cognitive shortcut. Human minds are excellent at recognizing agents: faces, intentions, plans, leaders. When a system behaves coherently, we reach for the story of a coherent self. That story is often wrong, and the wrongness matters.

Start with the royal language that clings to insects: queens, workers, soldiers. It suggests a court. It suggests orders. But a queen ant does not *rule* in the human sense; she is better described as a reproductive organ the colony protects. An ant colony is not an army with a general; it is a chemical economy with legs. The more carefully one looks, the less the metaphors of command help. What looks like obedience is often simply local rule-following under strong constraints. What looks like deliberation is often an accumulation of tiny, biased interactions that stabilize into a collective decision.

Artificial swarms inherit the same misunderstanding, only with higher stakes because we build them for our purposes. If a robot swarm maps a disaster site, it is tempting to imagine a "swarm mind" that knows the building. But knowledge is not evenly spread through the group. Some robots may have excellent local maps; others may have nothing but a few recent headings and a fading radio link. The pattern that looks like a plan can be a side effect of crowding, signal strength, or a tuning parameter chosen in simulation. The swarm can be competent without being self-aware, and it can fail catastrophically without intending anything.

The challenge, then, is to take swarm cognition seriously without pretending it is simply human cognition distributed across many bodies. The phrase *mind without a face* is useful because it blocks the most seductive mistake. A face implies a point of view. A face implies a single stream of experience, a place where "what the system believes" becomes a meaningful sentence. A swarm rarely has that. If anything deserves the title "mind," it is the pattern of interaction itself: a web of partial signals, local memories, and environmental traces that, in combination, can solve problems no individual could solve alone.

One of the clearest ways to make this idea respectable is to remember that humans already rely on cognition that is not confined to a skull. Edwin Hutchins's work on distributed cognition, culminating in his 1995 study of navigation teams, treated a ship's bridge as a cognitive system: people, instruments, procedures, and shared representations working together to produce a result no single person could produce. No one imagines the ship has a "self" in the way a sailor does, yet it is perfectly sensible to say that the navigation team-plus-tools performs computation. The "mind" in this account is not a ghostly overseer; it is a pattern of coordinated activity that spans brains and artifacts.

Swarm systems are similar, only more alien. In a navigation team, people can talk, explain, and correct one another. In an ant colony, the shared representation is often chemical. In a robot swarm, it might be a map that exists nowhere in full: fragments on different processors, a few key landmarks stored in a base station, a trail of digital breadcrumbs left in the environment. The cognitive work is real, but it is not housed in a single narrative stream.

This is where anthropomorphism becomes actively dangerous: it encourages a search for leaders and intentions when what matters are feedback loops and constraints. A human commander can be reasoned with, blamed, persuaded, trained. A feedback loop cannot. If a swarm gets stuck in a self-reinforcing pattern, the right question is not "why did it choose that?" but "what local evidence made that pattern stable, and where could a brake be applied?"

The discomfort deepens when the swarm is competent. Consider how quickly people attribute *understanding* to systems that coordinate well. A flock of drones maintaining formation through wind gusts looks as if it is "trying" to stay together. In reality, each drone may be running something closer to a reflex: maintain a target distance, adjust heading, avoid collisions, repeat. The competence is emergent, and emergence invites a particular kind of illusion: because the group-level pattern is coherent, we assume group-level comprehension.

There is a second illusion, almost the opposite: because no single unit knows the whole, people assume no one knows anything worth calling knowledge. That is also wrong. Swarms can encode information in ways that are not obvious to introspection. A pheromone trail is a memory written onto the ground. A robot swarm's spatial distribution can be a memory: high density here means "we have been drawn here recently," low density there means "we are neglecting that region." A temporary communication backbone is a memory of where information is needed most. None of this requires a central map, but it is still information-bearing structure.

Once swarm cognition is treated as real-but-nonhuman, a set of practical questions becomes clearer.

One concerns transparency. If a person makes a decision, we can ask for reasons, even if the reasons are imperfect. If a centralized algorithm makes a decision, we can often inspect code, logs, or model outputs. With swarms, the "reason" for a collective action may be distributed across thousands of micro-decisions and a changing environment. The causal story can be long, branching, and sensitive to tiny perturbations. This is not merely inconvenient for scientists; it is a problem for anyone who must trust the system under pressure.

Engineers and policy makers have begun to borrow the language of trustworthy technology to face this. The European Commission's High-Level Expert Group on AI (2019) framed trustworthy AI around ideas such as robustness and safety, transparency, accountability, and human agency. Those terms were not invented for swarms, but they apply with a twist. Transparency cannot mean "explain the decision of the decision-maker" if there is no decision-maker. Accountability cannot mean "identify the responsible agent" if the agent is a pattern of interaction. Robustness cannot mean "prove the component is correct" if the failure arises from components behaving correctly together in an unlucky context.

This mismatch produces a familiar human response: we reach again for hierarchy. If the swarm is opaque, give it a supervisor. If responsibility is diffuse, assign a commander. If behavior is emergent, restrict it with top-down rules. Sometimes this is sensible. Often it simply relocates the problem. A "human in the loop" who can only occasionally communicate with a swarm is not truly in the loop; they are a spectator with veto power, and veto power arrives too late if the system moves faster than deliberation.

A more honest approach begins with the idea of *assurance*: building structured arguments, supported by testing and design constraints, that a system will behave within acceptable limits. Recent work on assurance for emergent robot swarms (for example, frameworks discussed under labels like AERoS in the early 2020s) tries to make this practical by focusing on interactions: what messages can be sent, how local rules compose, what invariants can be enforced even as global patterns shift. The goal is not to remove emergence, which would remove much of the swarm's value, but to shape the space in which emergence is allowed to occur.

Security pushes the point further. A centralized system has obvious vulnerabilities: compromise the hub, and you own the system. Swarms trade that vulnerability for subtler ones. A hostile agent does not need to break every robot; it may only need to bias local cues so that the collective drifts. A single "rogue" unit can sometimes have outsized influence if local rules treat neighbors as trustworthy by default. Work by researchers such as Chen and Ng (2021) has explored lightweight ways to detect or limit rogue influence in swarms, precisely because heavyweight cryptography or constant authentication can be too costly for small robots with limited energy and bandwidth.

The deeper lesson is that swarms are neither inherently safer nor inherently more dangerous than centralized systems; they are dangerous in different ways. They are resilient to random failure but can be fragile to structured manipulation. They are adaptable in unfamiliar terrain but hard to predict in corner cases. They can continue functioning under partial communication loss, yet that same partial communication can make it difficult to intervene when things go wrong.

These technical facts feed directly into ethical ones. When people discuss autonomous weapon systems, they often speak of "accountability gaps": the worry that, when harm occurs, no one can be held responsible in a meaningful way, because causality is distributed across designers, operators, commanders, and the machine's own adaptive behavior. Philosophers and legal scholars have mapped these gaps into layers—technical, sociotechnical, governance—and have proposed forms of comprehensive human oversight. Even outside warfare, the same structure appears anywhere a swarm can affect human lives: policing, border surveillance, crowd control, industrial safety, emergency response.

The distinctive ethical unease with swarms is not that they are "mindless." It is that they are *mind-like* in their effects while remaining *personless* in their operation. A swarm can allocate attention, prioritize regions, persist in a behavior, and resist disruption. These are the kinds of verbs we use for agents. But if we treat the swarm as an agent, we risk granting it a moral status it does not have, or worse, using it as a moral shield: "the swarm decided," "the system concluded," "the drones thought."

A concrete historical analogy helps here: the early days of industrial accidents in complex factories. When a machine injured a worker, it was tempting to blame the worker's carelessness or the machine's "malfunction." Over time, safety engineering reframed accidents as system outcomes: guard rails, training, procedures, maintenance schedules, incentives, and workplace culture. Responsibility became less about a single guilty moment and more about a network of design choices. Swarm systems invite a similar reframing, but with additional complexity because the "machine" is not one artifact; it is many interacting artifacts, plus the environment, plus human operators who may only partially understand what is happening in real time.

This is the point at which the language of faces and rulers becomes not only inaccurate but evasive. If one calls a bee colony "obedient," one can ignore the chemistry. If one calls a robot swarm "autonomous," one can ignore the design decisions that shaped its autonomy. If one calls the system "intelligent," one can ignore the narrowness of its competence and the sharp edges of its failure modes.

A better habit is to ask what kind of intelligence is actually present. Swarm cognition is often *ecological*: it is inseparable from the environment in which it unfolds. Ant trails are not just internal signals; they are literally written into the world. Robot swarms that use beacons, dropped markers, or communication chains are doing the same. Intelligence here is not a private mental state but a pattern of coupling between agents and surroundings. That coupling can be powerful and also, as the ant mill shows, dangerously self-reinforcing when the environment feeds the system misleading evidence.

The practical benefit of respecting swarm cognition on its own terms is that it changes how one designs and governs these systems. Instead of demanding a single explanation for a collective action, one can demand logs of local interactions and tools for replay. Instead of asking for a single "decision trace," one can ask for bounds: guarantees about spacing, about maximum speed near humans, about how quickly a swarm must disperse when congestion rises. Instead of assigning all responsibility to an operator who may lack real control, one can distribute obligations across designers (who choose local rules), deployers (who choose contexts of use), and monitors (who watch for early signs of pathological feedback).

It also changes how one speaks to the public. People are understandably alarmed by the thought of "a mind of drones" overhead, because minds are associated with intentions and intentions with threat. The reality is both less and more unsettling. Less, because the swarm may have no inner narrative, no malice, no ambition. More, because absence of malice does not imply safety. A distributed system can do harm through the ordinary operation of its rules, especially if those rules have been optimized for efficiency in a narrow sense.

To make this contrast vivid, it helps to lay out the difference between hierarchical intelligence and swarm intelligence without implying that one is superior in general. The point is category, not ranking.

Human/Hierarchical intelligence (typical)	*Swarm/Distributed intelligence (typical)*
A central point of view; decisions can often be narrated as reasons.	No single point of view; "reasons" are patterns of local interaction and feedback.
Explicit plans and symbols (maps, rules, goals) held in memory.	Implicit plans embedded in local rules; memory often stored in the environment (traces, markers, spatial distribution).
Failure often tied to key nodes (leaders, hubs, critical components).	Resilient to random loss of individuals; vulnerable to interaction traps and structured manipulation of local cues.
Accountability can be attached to identifiable decision-makers.	Accountability must be assigned across design, deployment, monitoring, and governance because causality is distributed.
Learning often involves internal model-building and abstraction.	Adaptation often involves tuning thresholds and feedback; learning can be collective without any unit "understanding."

Respecting the mind without a face means learning to be precise about what one is admiring and what one is fearing. A swarm can be brilliant at search and terrible at explanation. It can be robust to broken parts and brittle to the wrong signal. It can be ethically appealing in rescue because it degrades gracefully, and ethically alarming in surveillance because it can persist and spread with little visible locus of control.

There is a final discipline, more personal than technical. It is the discipline of refusing the comfort of familiar stories. When a swarm behaves well, the familiar story is that it has a leader, a plan, a competence like ours. When it behaves badly, the familiar story is that it "malfunctioned" as if a moral agent had sinned. The harder truth is that the swarm's intelligence is real, but it is a different kind of real: not a self behind the eyes, but a moving pattern in which simple parts, nudged by local signals, can produce outcomes that deserve both awe and caution.

A humane relationship with such systems begins by granting them neither dignity nor blame, but something more useful: careful boundaries, honest language, and designs that assume confusion will occur and must be survivable. A swarm does not need a face for us to be responsible for what it does.

www.ingramcontent.com/pod-product-compliance
Lightning Source LLC
LaVergne TN
LVHW022212200926
848481LV00006B/1936